Emergent Programmatic Form-ation

Yehia Madkour
Oliver Neumann

Emergent Programmatic Form-
ation

Parametric Design Beyond Complex
Geometries

VDM Verlag Dr. Müller

Impressum/Imprint (nur für Deutschland/ only for Germany)

Bibliografische Information der Deutschen Nationalbibliothek: Die Deutsche Nationalbibliothek verzeichnet diese Publikation in der Deutschen Nationalbibliografie; detaillierte bibliografische Daten sind im Internet über http://dnb.d-nb.de abrufbar.

Alle in diesem Buch genannten Marken und Produktnamen unterliegen warenzeichen-, marken- oder patentrechtlichem Schutz bzw. sind Warenzeichen oder eingetragene Warenzeichen der jeweiligen Inhaber. Die Wiedergabe von Marken, Produktnamen, Gebrauchsnamen, Handelsnamen, Warenbezeichnungen u.s.w. in diesem Werk berechtigt auch ohne besondere Kennzeichnung nicht zu der Annahme, dass solche Namen im Sinne der Warenzeichen- und Markenschutzgesetzgebung als frei zu betrachten wären und daher von jedermann benutzt werden dürften.

Coverbild: www.purestockx.com

Verlag: VDM Verlag Dr. Müller Aktiengesellschaft & Co. KG
Dudweiler Landstr. 99, 66123 Saarbrücken, Deutschland
Telefon +49 681 9100-698, Telefax +49 681 9100-988, Email: info@vdm-verlag.de

Herstellung in Deutschland:
Schaltungsdienst Lange o.H.G., Berlin
Books on Demand GmbH, Norderstedt
Reha GmbH, Saarbrücken
Amazon Distribution GmbH, Leipzig
ISBN: 978-3-639-19930-7

Imprint (only for USA, GB)

Bibliographic information published by the Deutsche Nationalbibliothek: The Deutsche Nationalbibliothek lists this publication in the Deutsche Nationalbibliografie; detailed bibliographic data are available in the Internet at http://dnb.d-nb.de .

Any brand names and product names mentioned in this book are subject to trademark, brand or patent protection and are trademarks or registered trademarks of their respective holders. The use of brand names, product names, common names, trade names, product descriptions etc. even without a particular marking in this works is in no way to be construed to mean that such names may be regarded as unrestricted in respect of trademark and brand protection legislation and could thus be used by anyone.

Cover image: www.purestockx.com

Publisher:
VDM Verlag Dr. Müller Aktiengesellschaft & Co. KG
Dudweiler Landstr. 99, 66123 Saarbrücken, Germany
Phone +49 681 9100-698, Fax +49 681 9100-988, Email: info@vdm-publishing.com

Printed in the U.S.A.
Printed in the U.K. by (see last page)
ISBN: 978-3-639-19930-7

Contents

List of Figures

Preface

Emergent Programmatic Form-ation is based on a thesis project by Yehia Madkour conducted with Oliver Neumann (thesis chair) in the Master of Advanced Studies in Architecture program at the School of Architecture and Landscape Architecture at the University of British Columbia. Halil Erhan from the School of Interactive Arts and Technology at Simon Fraser University and Matthew Soules from the School of Architecture and Landscape Architecture at the University of British Columbia provided valuable input as committee members for the thesis studies. The thesis research has been edited and reformatted for this publication.

Through a series of design studies at varying scales, *Emergent Programmatic Form-ation* illustrates how parametric modeling can be productively integrated into current design and planning processes to coordinate complex relationships and conditions specific to urban housing.

Definitions

Responsive: The ability to dynamically respond (react) to different criteria in the design problem.

Computation: Any type of information processing that can be represented by a sequence of operations. This includes phenomena ranging from human (logical) thinking to mathematical calculations.

Algorithm: A set of well-defined rules (procedure) for the solution of a problem in a finite number of steps.

System: A set of interacting or interdependent elements designed to work as a coherent entity.

1. Introduction

The increasing use of digital techniques in design clearly exhibits the instant and inevitable temptation to develop complex formal solutions. However, the parameters that influence form-making decisions are often too abstracted to address specific references. Although parametric modeling holds rich potential for controlling multiple processes and transformations, it is a bigger challenge and responsibility to employ digital design techniques in designs responsive to particular conditions of context, user functions, and program that are based on a complex set of interrelated natural and social references.

Emergent Programmatic Form-ation illustrates how parametric modeling can be productively integrated into the current design and planning process to coordinate complex relationships and conditions specific to urban housing.

Chapter 1 reviews key theoretical approaches that analyse how natural and artificial systems operate. In particular, the chapter provides the conceptual background and methodology that defines the computational and parametric approach to the design proposed and applied throughout the thesis.

Chapter 2 presents three existing computational design techniques and related precedents. The chapter formulates the parametric design approach for the research and proposes a shift in the digital design process away from formal ambiguity to spatial configurations that reflect a complex array of programmatic performance criteria.

Chapters 3 to 6 present a series of parametric design explorations. The explorations define new grounds where digital techniques are applied to manipulate not only form, but also function, program and space. In each of the explorations, dynamic relationships are created between the model and its influencing programmatic factors. In the studies, the digital model stores explicit performance-sensitive design decisions and constraints and responds to them simultaneously. This allows for a complex and cumulative expression, both programmatic and spatial:

Chapter 3: *Space Configurations* uses social and environmental criteria to generate a wide range of spatial configurations and modes of occupation within a typical city apartment;

Chapter 4: *Tower Formation* explores how the physical and environmental context affects the design formation of a residential tower in Vancouver's downtown;

Chapter 5: *Shade* is a responsive sun shading system that calculates its cardinal orientation and accordingly controls the lengths of horizontal and vertical overhangs.

2

Chapter 6: *City Configurations* engages the city at a larger scale and explores a responsive way of planning and regulating the relationships between buildings in the city by establishing a dynamic base that incorporates a complex set of city planning guidelines.

Chapter 7 presents conclusions derived from the explorations of the previous chapters.

This research project focuses on the integration of computation into the design process. Computational design techniques are applied to a series of explorations of varying program and scale to establish architecture that is responsive to programmatic and context sensitive criteria through the design process.

Computation and digital techniques can help inform the design process. Computational and parametric design systems respond to varying pragmatic criteria that address a complex set of interrelated natural, social, and economic references. In this context, it is significant that parametric modeling holds rich potential for controlling multiple processes and transformations. The main goal of this research study is to explore the potential of parametric modeling to dynamically resolve and control the particular conditions of context, user functions, and program.

Related parameters specific to context, project program, and user-sensitive design are explored within the digital design field. In a series of design explorations, form making is influenced by and responsive to programmatic and functional issues arising from each exploration's design problem. A study of relevant architectural design precedents reveals a prominent disregard of program-sensitive approaches to design problems. To date, the drivers that influence form-making decisions are often too abstracted to address programmatic design issues. This research project explore a more challenging approach of designing spaces considering issues related to user functions and project programs, User-oriented programmatic functions hold many limitations for the production of form. These limitations- accompanied by the still limited degree of control associated with algorithmic techniques- explain the general disregard of program-sensitive design approaches in the field.

Urban Housing

In this study, housing design in metropolitan areas is used as a platform for investigating computational techniques in the architectural design process. The research explores parametric modeling techniques to incorporate and coordinate complex relationships and conditions specific to urban housing. User-oriented programmatic functions based on required performance criteria, typical standard sizes and proportions of domestic spaces, efficiency, user preferences, as well as the natural and artificial contexts and their

3

influence on buildings, all hold limitations for the production of form. In the proposed parametric design systems, the multiple design criteria and limitations are used as the design drivers and constraints on which the systems operate and to which they respond.

This research puts the common use of prototyping to generate architectural form in question. Distinctions between user needs and ways of living and interaction correlate to spatial differentiation and variation.

The technique to design countless varieties of housing options is obviously different from that needed to produce fixed prototypes. This project explores the capacity of parametric modeling to coordinate a wide range of pragmatic parameters. The study illustrates that manipulating multiple processes via a computational design is the method to avoid architecture prototyping and to better address interrelated needs.

Methodology

This study establishes its position between design and research by integrating phases of research throughout the design process. Initially, the project focuses on historical research and background theories that inform computational design approaches. This discussion is illustrated by case studies of relevant design projects. Subsequently, a series of design explorations investigate the capacity of parametric design to coordinate a wide range of performance-related design parameters in an architectural design process. Throughout each of the explorations, action research and reflective assessment support the planning and application of computational design systems. The study progresses through consecutive phases of research and explorations that produce design systems that are dynamically responsive to their functional context, thereby informing design studies throughout the process.

History: The initial historical research focuses on an overview of key theories significant to digital technologies and their implications for architectural design. Complexity, Emergence, and Natural Systems constitute key reference points for the historical theory overview. These concepts are then used to formulate the proposed design approach.

A study of digital design techniques used in architectural projects reveals the evolution of applications of computers in the design of architecture. The selection of parametric modeling as design techniques is based on an understanding of existing digital design methods and their potential and limitations.

Case Studies: The case studies explore prominent architectural precedents that utilize digital systems in the process of form generation, revealing a general temptation to focus on formal solutions and to disregard function and program sensitive approaches.

4

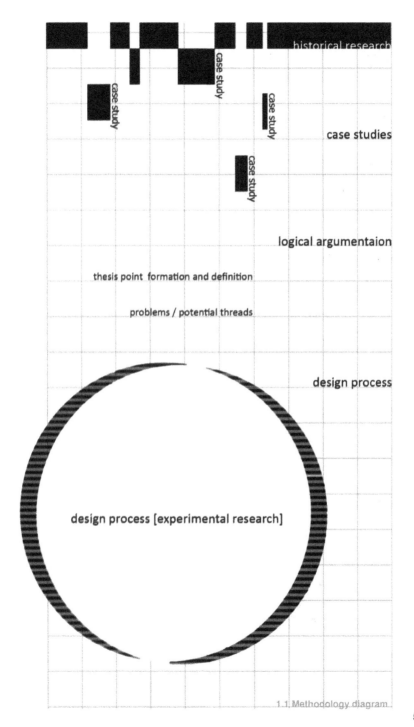

historical research

case study

case study

case study

case studies

case study

logical argumentaion

thesis point formation and definition

problems / potential threads

design process

design process [experimental research]

1.1 Methodology diagram

5

Design process: Apparent considerations that figure into the design of residential architecture in the city structure the design research. The research is undertaken through a series of design explorations aimed at identifying the potential of digital design processes. Through a series of design explorations, action research and reflective assessment inform the experiments. The computational design systems proposed in each of the explorations incorporate relevant design constraints into the design model, thereby creating a dynamic link of reflective assessment. Subsequent design explorations are informed by previous design phases.

Theoretical References

The computational approach to this research is based on an understanding of natural and artificial systems that constitute our surroundings.

Many of the current design approaches remain committed to outdated conceptual references and models. Acknowledging how technology has dramatically transformed social interactions, has affected global political developments, and has introduced new approaches to space, it is possible to develop approaches to space and design that correspond to the complexity and dynamism of natural, cultural, political, and economical processes. The way we live and experience these adaptive systems prompts us to change the way we design.

Theories of complexity and emergence are the key references for a conceptual understanding of the surrounding adaptive systems. They also serve as inspiration for the computational design processes in this study.

Complexity

The world around us exhibits complexity at all levels, from the interdependencies of financial markets and political processes to the ecologies of living organisms. Each system is highly organised and governed by the interaction between its elements. Complex systems result from the relationship between the individual elements, their collective behaviour as a consequence of their individual behaviour, and the systems' responsiveness to environmental parameters. Characteristic for complex systems is behaviour that is not obvious from the properties of the individual parts. Special cases of complex systems are complex adaptive systems (CAS); they are adaptive in that they have the capacity to change and learn from experience.

A study of urban centers reveals places marked by interconnected natural, cultural, political, and economical processes. These interactions contribute to a complex system

6

that, although controlled by laws and regulations, displays features of spontaneous self-organization.

Static Models: Studying complex systems heavily relies on 'reduction'. Applied to architectural design, a scaled model of a building reduces a design problem to its essential information. Contributing parameters can subsequently be more easily comprehended and manipulated. This method works for conventional design as well as for computational approaches. A model serves to simplify the building and highlight the important elements relevant to the design in order to inform design decision making while being conscious of the interactions between the design features.

Emergence

> "We are everywhere confronted with emergence in complex adaptive systems- ant colonies, networks of neurons, the immune system, the internet, and the global economy, to name a few - where the behaviour of the whole is much more complex than the behaviour of the part." (Holland 2).

Concepts of emergence provide a productive description of complex systems. The above quote is from the opening chapter of John Holland's 'Emergence from Chaos to Order'. Holland's work is very effective in defining technical concepts as a foundation for studying emergence.

A basic definition of emergence- *much coming from little-* applies to board games such as Tic Tac Toe. In these games, complexity results from the application of simple rules or laws. While a number of basic rules govern the development of the board game, the arrangements of pieces on the board grow into a more complex state with every round of play. It is impossible to understand the state of a game in progress by understanding the state of individual players on the board. The players, guided by the rules of the game, interact to support one another and to control various parts of the game. Power relationships continuously change the state of the game. The game quickly rises to a complex level. The degree of complexity changes over time although the guiding rules are invariant.

Dynamic Models: In contrast to models with static form, such as conventional scaled architecture models, dynamic models change in configuration. The key in constructing dynamic models is finding fixed laws that generate the configuration change and define the interaction between smaller components of the model. These laws correspond roughly to the rules of a game that define permitted ways of moving or placing the pieces on the board. Using this logic, we can construct models that exhibit dynamic emergent phenomena.

Natural Systems: Growth in natural systems can be understood in terms of complexity and emergence. Plant seeds contain specifications that produce complex and distinctive plant structures. A seed encapsulates the genes that specify step-by-step unfolding of biochemical interactions. An organism's inheritable information, its genotype, not only governs the way the genes interact with its surrounding in order for the plant to grow and survive but also affects its phenotype, its outward physical manifestation.

Una-May O'Reilly, Martin Hemberg, and Achim Menges use references to genotypes and phenotypes to describe the complex process of natural growth. O'Reilly, Hemberg, and Menges explore the potential of generative algorithms as operative design tools. Their exploration is intended "... to instrumentalise the natural processes of evolution and growth, to model essential features of emergence, and then to combine these within a computational framework" (O'Reilly 49) with the aim to develop a generative design tool that can produce complex architectural forms.

The concept of emergence provides both an explanation of how natural systems evolve as well as a process for the creation of artificial systems that model complex behaviour. Gilles Deleuze suggested that the capacity for morphogenesis is inherent to matter. It is a result of an ongoing exchange between a system and its environment[i].

Frei Otto's form-finding experiments to resolve lightweight structures are examples of material efficiencies in complex structural arrangements. In 'Frei Otto in Conversation with the Emergence and Design Group', Otto discusses the development of modelling through form-finding techniques inspired by natural systems. Structural characteristics of soap bubbles and spider webs provide reference for efficient lightweight structures and demonstrate how form can be generated by analysing natural systems.

Emergence in architecture: Emergence in architecture implies the creation of artificial systems designed to produce forms that exhibit complex behaviour. Current software applications and computers help to model emergent systems. The processing power of computers offers the possibility to compute a range of interrelated rules and constraints and to dynamically represent the accumulated effects. As a result, architecture, though typically considered a static art, can become a part of the complex adaptive systems of its surroundings.

Architects need to rethink how architecture is designed; they need to develop a methodology that takes into account a design environment in flux. Concepts of emergence applied to architecture allow thinking of a building as a complex set of relationships with the design evolving to higher degrees of complexity and responsiveness.

It is not necessary that a building itself is viewed like the living organism, but it is the design process that has to evolve. Helen Castle, in 'Emergence in Architecture', calls for constructing architecture that has evolved through a process of morphogenesis, rather than regarding buildings as unchanging, isolated tectonic objects.

> *'Emergence requires the recognition of buildings not as single bodies, but as complex energy and material systems that have a life span, and exist as part of the environment of other buildings, and as an iteration of a long series that proceeds by evolutionary development towards an intelligent ecosystem'.* (O'Reilly 51)

Emergent systems in architecture are needed to introduce responsiveness into the design process. The introduced dynamic relationships help to understand how the interaction of varying social, cultural, and economical parameters can affect spaces.

Applying the concept of emergence relies on blurring the boundaries of once separate sciences. Overlapping domains of developmental biology, physical chemistry, mathematics, and computer science are central references for the application of concepts of emergence to architecture. The convergent lines of thought between biology and mathematics were initiated early in the 20th century, particularly in the work of D'Arcy Thompson. D'Arcy Thompson regarded the material forms of living organisms as diagrams of the contextual forces they are exposed to[ii]. In 'Digital Morphogenesis', Neil Leach encourages the use of theoretical and scientific work from other disciplines to promote concepts of morphogenetic architectural design[iii]. Leach believes that Gilles Deleuze's theories on material behaviour and D'Arcy Thompson's theory of transformation will inform architecture practice to incorporate complex relationships as parameters in the design process.

Design Approach

The design projects explored in this research study use rule-based procedures to generate complexity out of simple initial states.

Rule-based design is the key to building dynamic models that respond to changing influences. The systematic unfolding of programmed parameters and relationships allows for continuous recalculation of the building model. The responsive and adaptive character of such a system makes it relatively easy to examine the effects of manipulations. This design approach allows responsive relationships to exist between initial design parameters and the resulting spaces.

Subjectivity: The application of concepts of emergence into the architectural design process has its limitations. A design process that exclusively focuses on scientific logic does not allow considering subjective design decisions that are central to designing in a cultural context. Artistic sensibility and creativity of architects are often seen as incompatible with the integration of computer algorithm and scientific logics into the design process. However, concepts of emergence and subjective and culturally sensitive decisions can coexist in the design process through the incorporation of selective parameters into the design process. Designers set the parameters for the design and maintain the ability to interpret and control the formation process, thereby allowing subjective decisions and rule-based design processes to coexist.

---

[i] DeLanda, Manuel. "Deleuze and the Use of Genetic Algorithms in Architecture." AD Wiley Academy January 2002: 9-12.

[ii] Thompson, D'Arcy Wentworth. On Growth and Form. Cambridge University Press, 1942.

[iii] Leach, Neil. "Digital Morphogenesis: A New Paradigmatic Shift in Architecture." Archithese 2006: 44-49.

2. Design Technique

Computers in Architecture

The application of a rule-based design approach requires vigorous processing power with the potential to execute complex procedures. Computers are a necessary tool for the realisation of such design approaches. However, in the architectural design process, computers should contribute to the design process, and not simply be used as sophisticated drafting tools. Currently, the majority of architects design by conceiving spatial forms after the consideration of program, context, and other design-influencing factors. These considerations reflect the designers' conception and vision and are ultimately translated into building models. Plans, elevations, and sections are examples of building models, as are renderings and scaled three-dimensional models. Used in such a way, computer software packages are limited to the production of representations and visualisations of the conceived spaces. The link between the designed spaces and initial influencing factors relies entirely on the designer's ability for abstraction.

By incorporating parametric modeling, it is possible to include design influences and constraints as active parameters into the design process. As a result, dynamic relationships related to concepts of emergence that allow for designs responsive to complex behaviours can become integral parts of the design process.

> *'We can use the mathematical models ... for generating designs, evolving forms and structures in morphogenetic processes within computational environments'* (Weinstock 17)

Abstraction is a key element in generating building models. A system has to be reduced to its essential information to allow for its manipulation. In parametric systems, numbers are used to describe and control the model's information such as dimensions, coordinates, and angles.

Digital Design Techniques

The review of case studies that incorporate three existing computational design techniques in the design process helped to formulate the parametric design approach of this research project. The findings of the case study review suggest a shift in the digital design process away from formal ambiguity to spatial configurations that reflect programmatic references. The three distinct computational design techniques used in the case studies share a rule-based approach to design. Rule-based design approaches rely on analysing the purpose of design, abstracting the design approach to a set of rules, and setting a systematic procedure for a design to emerge.

1. Analog Computing Processes

Key Characteristics:

1. Abstraction

2. Rule setting logic

3. Manual execution of rules

4. Use of physical models in the process of design

The analog computing process of design demonstrates cases where a design approach is abstracted to a set of rules to be followed in an analog environment. This process is illustrated in the form-finding experiments of Frei Otto to resolve lightweight structures. Frei Otto's work demonstrates how forms can be generated based on studies of growth in natural systems. The analysis of a system's response to surrounding forces and the identification of laws that govern these responses is a rule-based approach to design that does not necessarily rely on computer software. However, adhering to rule setting logics is central to form generation.

Case study 1:

Title: Son-O-House

Date: 2004

Location: Son En Breughel, The Netherlands

Program: Sculptural pavilion for interactive sound

Architect: Lars Spuybroek (NOX)

2.1 Son O House

Overview

The Son-O-House structure is both an architectural and sound installation that allows visitors to participate in the generation of the sound. The sound work, made by composer Edwin Van Der Heide, is continuously generating new sound patterns that are activated by sensors and triggered by movements of visitors.

Lars Spuybroek describes this pavilion as 'a house where sounds live' (Spuybroek 174) and as a structure that refers to living and body movements that accompany habit and habitation. Departing from this point, the design of the structure is derived from a network of large- and small-scale body movements. The movements are translated into an intricate configuration of intertwining lines that are equally a reading of movements at various body scales as well as the basis for a structure. The complex surfaces resulting from the arrangement of the lines are translated into a series of intersecting flat ribs that support the outer skin; double-curved surface tiles form the cladding of the structure.

Process

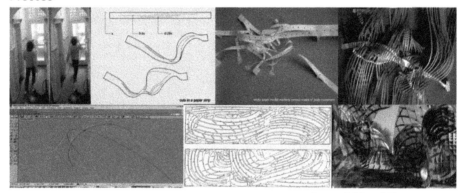

2.2 Son O House process

Lars Spuybroek used analog rule-base design techniques to represent movement at various body scales and to determine the structure Son-O-House. The design process can be described in 5 phases:

1. A camera is set to record human movements at different scales. The movements include walking as well as movements related to activities at desks and sinks. (Figure 2.2)

2. The initial mapping of the body movements led to model studies in paper. Insertions into paper bands at varying lengths correspond to types of movement: 'an uncut area

14

corresponds with the bodily movement, a first cut through the middle corresponds with limbs, and finer cuts correspond with movements of the hands and feet' (Spuybroek 174).

3. White paper bands are twisted into various configurations. The pre-informed paper paths are connected at points with most connective potential. As a result, curvature emerges.

4. Spatial volumes are generated by sets of red parallel paper strips bent over and through the white path strips in perpendicular direction. Form emerges as result of the reactions of the materials that are exposed to bending and twisting forces.

5. The paper models are then digitized, followed by a computer-aided analysis of the forces and stresses of the resulting complex geometries. Shapes of structural components are then derived from the refined models for the digital fabrication process.

Comments

Son-O-House is an example of an analog computing approach to design. While it is clear how the initial design approach is abstracted into a number of rules for manual processing, the design process lacks responsiveness. It is not open to changing references and rules. In addition, the manual execution of design rules is time consuming and limited to basic relationships and dependencies of design parameters.

2. Scripted Algorithms

Key Characteristics:

1. Abstraction

2. Rule setting logic

3. Translation into defined steps

4. Execution in a digital computational environment

Designs based on scripted algorithms are the most explicit in the way a design process is deduced from a procedure of rules. Generally, an algorithm is a set of rules for the solution of a mathematical problem. A rule-based approach to design corresponds to scripted algorithms. Well-defined instructions are carried out in a digital environment with the help of the processing capacities of computers. A very powerful attribute of this technique is the use of conditionals and loops. Loop and conditionals are sequences of statements that may be carried out a specified number of times or until a particular condition is met. Looping allows for repetition in design with changing variables.

Case study 2:

Title: Log Cabin

Date: 2005

Location: Competition

Program: Log Cabin

Architect: Benjamin Aranda and Chris Lasch

Benjamin Aranda and Chris Lasch, as described in 'Tooling', follow a rule-based design approach by focusing on procedural thinking. Aranda and Lasch describe rule-based design as a recipe for shape generation. In their studies, they explore algorithms to simulate organizational principles such as packing, weaving, blending, and tiling.

Through a number of design research projects, Aranda and Lasch relate rules of organization to logics for construction. In the following case study, the designers investigate packing for a competition project to design a log cabin.

Process

1. The construction of a log cabin involves cutting trees and then stacking them. Aranda and Lasch investigated a reversed order to start their project. By first stacking and then cutting, they take advantage of the capacity of wood to influence the configuration of a structure by changing its shape from circle to ellipse to bar. Structural properties of the design and opacity become controllable through variations of the principles of stacking and packing.

2. A packing method controls the organisation of logs (Figure 2.3).
The step-by-step design approach for this project includes picking a random point, creating a shape of random size, and looping the process without intersecting the shapes. Although the applied packing recipe lacks a termination statement, its underlying algorithm is very powerful. Variations to individual forms and the related overall configuration can be easily manipulated by changes in the algorithm and its variables.

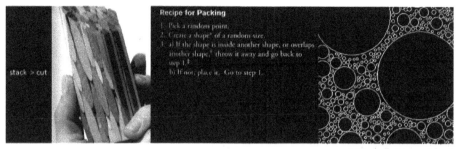

2.3 Log Cabin process 1

3. After designing a plan layout for the cabin, Aranda and Lasch applied the packing system for the logs with variations in the parameters of size, position, and repetition to the cabin's facades. (Figure 2.4)

2.4 Log Cabin process 2

Comments

Aranda and Lasch's project for a log cabin illustrates a strategic application of scripted algorithms in a design process. The project makes use of loops and the processing power of computers. A loop is a sequence of statements that is specified once but carried out a number of times until specific conditions are met. In the Log Cabin, project manipulations of geometrical parameters resulted in various spatial configurations of the logs.

However, an approach based on scripted algorithms does not allow for a dynamic adaption to changes. The design progresses on its preset path once the rules are set. As a result, the design can only be changed by deleting a previous model and by re-running the script.

17

3. Simulating Organic Growth

Key Characteristics:

1. Rule setting logic

2. Virtual environmental conditions

3. Population of designs

4. Fitness evaluation procedures

5. Execution in a digital computational environment

The design approach to this project uses generative algorithms to create complex spatial configurations. These configurations are generally inspired by growth of natural systems. Achim Menges uses parametric software to develop surface geometries with built-in modifying commands that simulate environmental conditions. This approach results in the population of adaptive geometries. Instances of the generated designs are evaluated and chosen according to a fitness function, which eliminates invalid geometries that do not fit the set criteria.

Case study 3:

Title: Pneumatic Strawberry Bar

Date: 2004

Location: London, England

Program: Strawberry Bar

Architect: Achim Menges

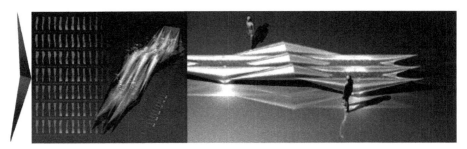

2.5 Pneumatic Strawberry Bar process

This project is a small-scale design for a strawberry bar for the Architecture Association annual exhibition. The design approach is based on the evaluation of geometries as a substitute for a structural analysis that would otherwise guarantee the structural integrity of the resulting folded form.

Process

1. The design is a folded structure based on two trapezoids connected at a common seem.

2. Using *Genr8*, a *Maya* plug-in, Achim Menges breeds a population of the trapezoidal elements according to geometrical fitness criteria that evaluate the target value and a fractional weight of the generated configurations.

These criteria include:
a. Symmetry
b. Size
c. Soft boundaries: to determine if the populated surfaces are allowed to grow through the boundary wall.
d. Subdivisions: to measure the amount of articulation in the surface.
e. Smoothness: to measure the extent of local geometry variations in the Z plane. Smoothness controls the shape of the base components.
f. Undulation: to measure global variations in the Z plane. This parameter controls the assembly of component population.

3. Changes of the fitness criteria dynamically change the population of the base component used to produce the structure's final configuration.

Comments

Achim Menges' design approach illustrates that design systems can be responsive. This approach allows for fast and precise interactions between the designer and the resulting geometries of the project. While the Pneumatic Strawberry Bar project demonstrates a useful integration of design decisions into the design process, the use of parameters pre-defined by software limits the designer's possibilities and restricts the processes of system growth.

Comparison of existing digital techniques

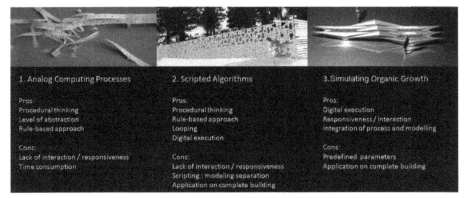

2.6 Comparison of existing digital techniques

Conclusion

All three identified digital design methods clearly illustrate how rule-based logics are used in design process for form generation. The case studies also illustrate the importance of abstraction in the design process in order to introduce rule-based procedures.

Rule-based design methods are central to responsive designs because they allow engaging in design as a process. While the analyzed design projects provide relevant references for the design research in this study, each of the project approaches has limitations that should be noted:

The analog computing process of the Son-O-House project allows for only limited engagement with dynamic relationships of design parameters due to the difficulty to process complex relationships simultaneously using analog methods. Digital design techniques are necessary to process complex conditions.

Both methods based on analog computing processes and on scripted algorithms lack the responsiveness needed to alter and manipulate existing designs throughout the design process. Both methods necessitate the designs to progress on a path of pre-defined rules without the capacity to adapt to changes in rules or parameters. In order to make changes to the design, scripts with adjusted code or guiding parameters have to be re-run.

Achim Menges' Pneumatic Strawberry Bar project offers a design approach that is responsive to changing parameters. Responsiveness is made possible by incorporating

20

design constraints into design process. Changes to any of the constraints result in a dynamic response in the design model.

Aranda and Lasch's design approach based on scripted algorithms allow for the integration of loops. Loops allow for the creation populations of a design typology, with slight variations occurring due to changing variables.

Breeding a population of designs can also be achieved by simulating organic growth similar to Achim Menges' design approach for the Pneumatic Strawberry Bar. In this example, however, the population is limited to predefined parameters that are inherent to the software used for the design.

The three design approaches identified in the case studies offer only limited options to incorporate dynamically changing parameters into the design process. In addition, responsiveness, although limited, primarily focuses on formal complexity without concerns for context, program, or functional considerations.

To date, the drivers that influence form-making decisions in digital design processes are often too abstracted to address programmatic design issues. With projects focusing on complex open space structures and surface geometries, user functions and project programs are generally not incorporated in the design process. This disregard of program-sensitive design approaches can be explained by the limited degree of control associated with common algorithmic techniques.

In contrast, the design studies of *Emergent Programmatic Form-ation* engage context and program specific as well as user-sensitive design drivers in a digital design process. This engagement is vital in order to respond to an extended framework of references that address the complex cultural context within which designs exist.

Parametric Design

In order to apply the context and program-sensitive design approach to a computational rule-based setting, a technique to coordinate and control multiple processes in a responsive and operable environment is essential. Parametric design techniques allow applying advantages of scripting techniques to a more interactive environment. Using parametric design tools, designers establish dynamic relationship between the contributing parameters of their designs. As a result, designers are able to manipulate, control, and interact with the design by adjusting parameters that affect the performative and formal characteristics of a project.

Parametric design techniques provide the designer with more control without the need to re-run scripts. Scripts developed by the designer define the geometric behavior of components within set dynamic relationships. Dependencies of design components are persistent throughout the model and can, therefore, be executed during dynamic modification.

Parametric modeling incorporates design decisions and constraints into the design and modeling process. Similar to digital models produced in CAD applications, parametric modeling is centered on geometric elements. However, parametric modeling offers the ability of store relationships between elements based on associations. This explicit storing of design rules coincides with a high level of abstraction. A design and its components are reduced to a network of interrelated geometries. The ability to display and interact with this dynamic network provides the designer with the ability to analyse and manipulate the design.

In order to establish a functional responsive design system, a high level of abstraction is necessary. Similar to functions guided by control panels, design problems need to be reduced to core factors in order to be modified. Parametric modeling uses numbers to manipulate complicated design systems. For example, *Revit*, a parametric modeller widely used in architecture offices, recalculates a building's geometries each time the architect manipulates dimensions of individual parameters. As responsive software, *Revit* is pre-programmed to understand certain number manipulations, it translates them into graphical representation of architecture, taking into consideration all the interacting relationships between floors, slabs, heights, etc. However, *Revit* is targeted towards the mainstream architecture production offices and is missing a fundamental aspect much needed to expand possibilities for designers. Limited by architectural conventions, the software lacks the capacity for expansion to incorporate additional users-defined features that would accommodate design-specific needs.

GenerativeComponents, in comparison, is a parametric modeler that allows for flexibility in the design exploration stage. Once the underlying logic and design relationships are defined, the designer can create new options and features that extend the system of dynamically related parameters beyond its pre-programmed features. This, among other advantages, allows for an iterative search for more efficient solutions.

> *"A key concept of GenerativeComponents is the ability to define geometry in different ways. One approach is based on traditional CAD modeling expanded by parametric associative aspects. A second approach is based on writing simple*

scripts that generate geometry and associations. A third one is through programming in the Microsoft Visual Studio environment". (Aish 1)

The three approaches to manipulating geometries exist in parallel and can be applied at different times towards the same system of design components. Designers without programming skills are given the opportunity to explore operations that would only be possible in scripting and programming. In addition, an assembly of pre-programmed features can be captured and turned into new features for future use, allowing designers to use the power of programming without the needs for programming expertise.

In contrast to CAD modeling programs, the investment is not primarily in geometries but in the relationships and dependencies that define how the geometries interact and respond to reference parameters.

In the next chapters, parametric modeling in *GenerativeComponents* is used in a series of design explorations to define and control dependencies between a design model and its programmatic, social, and environmental influences. Digital models are used to store explicit performance-sensitive design decisions and constraints in a setting that allows dynamic modification and response. The design explorations challenge the capacity of parametric design techniques to manipulate not only form, but also function, program, and space.

3. Space Configurations

Introduction

The first exploration applies parametric and computational techniques in design to generate varying spatial configurations for typical living spaces. The exploration works at the small scale of basic living units in a residential tower located in the dense urban center of Vancouver. The living spaces are defined by wall boundaries but are intended to respond to a wide range of living scenarios.

Design Approach

Space Configurations uses computational design techniques to generate varying spatial configurations inside a typical one-bedroom unit. Social and environmental parameters influence the configuration of an apartment to accommodate the life-styles and needs of possible occupants. Social factors such as lifestyle and specific needs of potential occupants together with environmental factors such as location, orientation, and view constitute parameters for the unit configuration. Parametric techniques are used to design for changing occupation by modeling the design components that are related to occupation constraints into one system.

Space

Space in the explored small high-rise living units is defined by its boundaries, physical and non-physical separations. Physical separations are the hard interfaces inside a given space such as walls, partitions, floors, and skin. Non-physical separations can be established through lighting, sound, and surface treatments. This exploration focuses on spatial conditions and explores physical separations of space and iterations of its components within the boundaries of a condominium unit defined by outer walls, floor, and ceiling to generate alternative living environments. Changes to the unit configuration are not intended to affect the overall tower structure.

Occupation

Occupation is defined by the activities performed within the space. Working, exercising, eating, partying, and sleeping all require different spatial characteristics. In a typical high-rise housing scenario, users are classified into categories according to the number of household members. Users often have to adapt to spatial configurations that best fit their needs but does not directly reflect their life style. This common method of categorizations does not correspond to the actual occupation of space. Potential users have a wide range of interests and needs, and particular requirements can dramatically change the inside configuration of their living units.

Existing Social Condition

The majority of housing units in downtown Vancouver are one-bedroom apartments. According to the 2001 census[iv], 82.6% of Vancouver's downtown dwellings are apartment units in high-rise (5+ storeys) buildings. This number highlights the residential density in Vancouver's city core that was in part promoted by recent conversions of office towers into residential buildings. Half the population in downtown Vancouver falls into the 20-39 age group, with the number of one-person households reaching 60.1% of all households. (Figure 3.1)

3.1 Existing social condition in downtown Vancouver

Typical Unit

Categorization: The residential towers in Vancouver are home to a variety of studio apartments and one, two, and three bedroom units. The majority of living units are one-bedrooms.

Open Concepts: The analysis of downtown Vancouver's typical unit layouts integrate open kitchens into living spaces to give the illusion of spacious living and to offer opportunities for interactions within the unit.

Efficiency: Space efficiency is a major focus of apartments in Vancouver: entrance areas and hallways are either reduced in size or completely eliminated and bathrooms are directly connected to the living area, in some cases with additional access to the bedroom. Bedrooms are directly accessed from the living area. (Figure 3.2)

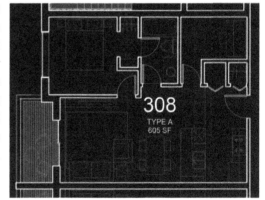

3.2 Typical living unit

26

View Exposure: A great emphasis is put on exposure to mountain and city views from living spaces and bedrooms.

Position

Unit layouts are highly standardised. This approach forces the potential users to adapt to the preconfigured spaces while life-style and activities vary from occupant to occupant.

Design system

A parametric system regulates the relationship between the sub-spaces and controls the physical separations inside the unit. In this system of dynamically related parameters that is developed using *Generative Components*, alterations at the level of one component affect other components in a generative and cumulative behaviour. As part of the rule-based procedures that are used to create a responsive design, system relationships between the design components are defined and their interactions are regulated. The configuration of each housing unit results from this interactive process.

Objective

1. The main goal for this design system is to develop space configurations that respond to a variety of social and environmental criteria.

2. This system should have the potential to act as a mass customisation tool for apartment models that controls and assembles the usual model components in ways to produce configurations for a range of users.

3. The system should also provide a dynamic base for designers to model new possibilities for occupation.

Design Process

Components of a typical unit:

1. Living area

2. Bedroom

3. Kitchen

4. Bathroom

5. Balcony

6. Entrance

3.3 Components of typical living units

Relationship between components

The floor layout of a typical unit can be understood in terms of relationships between subspaces of a living unit: kitchens are usually configured to be open to the living room; entrances lead directly into open kitchen spaces; bathrooms are often connected to both the living room and the bedroom; bedrooms are directly accessed from the main living space, and balconies are placed to connect to living space and bedroom. Living rooms and bedrooms are places to allow for mountain and city views.

These relationships of program elements are used in the parametric model to define and constraint the location of sub-spaces.

Parametric Model Unit

The typical living unit is modeled in *GenerativeComponents* to define and regulate the relationship between subspaces. The model components are the geometries that define the subspaces such as walls, floors, ceilings, etc.

Each unit layout is a variation of this base unit:

The boundary wall of the unit is fixed. The position of interior walls and floors (marked in white) and the glazing surface (grey) are controlled parametrically.

Defining Variables

1. Floors

Each of the floor components inside the model unit can either be:

I. Flat   II. Stepped   III. Ramped

To accommodate variations, variables are introduced to manipulate the heights at the each of the floor edges, the number of platforms and the height difference between platforms (value: 0 if flat), and the size of each floor component that affect room sizes.

2. Walls

Vertical separations between the unit components are achieved by partitions, which can also be controlled by variables. They can to be:

I. Full height   II. Percentage of full height   III. Non-existing

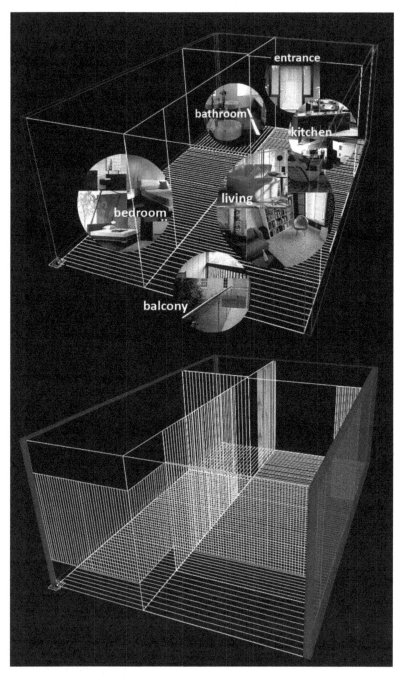

entrance

bathroom

kitchen

bedroom

living

balcony

3.4 Model unit

3. Skin

The exposed side of the unit is covered with a facade system that is manipulated to adjust the size and ratio of solid walls and glazing. Variables control the horizontal and vertical sizes of openings.

Variables Limits

Figure 3.5 shows the variables used for controlling the floors, walls, and skin components of the parametric model unit. The value shown in the table describes the initial configuration of the model.

input data	value	min/max
> Variable bed_living_In	0.4	0 —— 0.55
> Variable bed_living_Out	0.4	0 —— 0.55
> Variable depth_bed1	0.3	0 —— 0.5
> Variable depth_bed2	0.3	0 —— 0.5
> Variable h_in	zero	0 —— 0.2
> Variable h_out	zero	0 —— 0.2
> Variables wall_1	1.0	0 —— 1.0
> Variables wall_3	1.0	0 —— 1.0
> Variables wall_4	1.0	0 —— 1.0
> Variables wall_bath	1.0	0 —— 1.0
> Variable kitchen	0.2	0 —— 0.6
> Variable rise_living	zero	-1 —— 1.0
> Variable stairs_living	4.0	0 —— 4.0
> Variable rise_bed	zero	-1 —— 1.0
> Variable stairs_living	3.0	0 —— 3.0
> Variable terrace_depth	0.2	0 —— 0.2
> Variable horizontal	0.2	0 —— 0.5
> Variable vertical	zero	0 —— 0.5
> Variable hz_divisions	1.0	1 —— 4.0

3.5 Model variables

Upper and lower limits for these variables are set to control the desired range of configuration change for each of the components. For example, variables *bed_living_In* and *bed_living_Out* control the inner and outer floor separation between the living room and the bedroom. These two variables also control the position of the partition walls that separate the two spaces, controlled in height by *variables wall_1*, *variables wall_2*, *variables wall_3*.

The range of set values for the variables represents possible configurations. Beyond these values, the system's configurations are not valid.

Defining Constraints

Design constraints are embedded into the design model. This allows for the regulation of the design model behaviour and the related visual display of the design process. The

31

constraints are considered rules that control the interaction between components and that limit how the system's performance.

A list of constraints set possible values for the model variables:

1. The unit perimeter walls are fixed.
2. Same level access between the main living space, bedroom, and bathroom.
3. Living space and bedroom have outside exposure.
4. Balcony has direct access to living space.
5. Bedroom and bathroom have a common wall.
6. If ramps are included, floors do not exceed 1:20 inclination.
7. Height difference between platforms do not exceed 1ft.
8. Minimum dimension of bathroom: 20% of the unit depth.
9. Minimum width of living space: 30% of the unit width.
10. Minimum width of bedroom: 30% of unit width.

Design initiation

Social and environmental conditions are factors that influence the unit layout and performance of spaces. Social criteria- the social system- are occupant number, needs, lifestyle, and living preferences. Environmental criteria- the environmental system- are the unit's location within the building, its orientation and cardinal direction, and its exposure to view.

Social and environmental factors are directly related to performative aspects and the spatial configuration of a unit. (Figure 3.6) Manual adjustments to social and environmental factors trigger dynamic changes to the parametric spatial system.

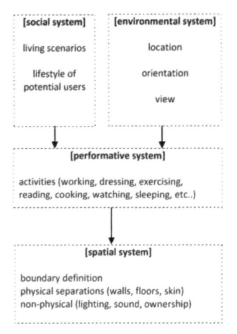

[social system]

living scenarios

lifestyle of potential users

[environmental system]

location

orientation

view

[performative system]

activities (working, dressing, exercising, reading, cooking, watching, sleeping, etc..)

[spatial system]

boundary definition
physical separations (walls, floors, skin)
non-physical (lighting, sound, ownership)

3.6 Design initiation

32

Social system

Census data for Vancouver provides the basis for possible households scenarios for one-bedroom apartments. Figure 3.7 describes possible number of occupants, gender, and age of members in one-bedroom households.

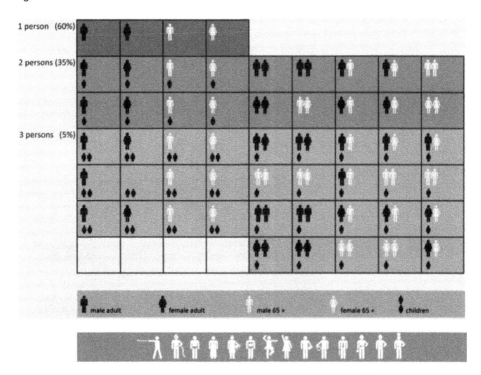

3.7 Possible households

33

Social factors that address accessibility, privacy, sociability, multi-functionality, number and type of occupancy, and desired level of exposure are identified in figure 3.8. These influencing factors correspond to parameters that control the geometries and components of the parametric model. Each of the criteria is associated with related values for the component variables. The resulting spatial configuration is illustrated in the output column of the schedule.

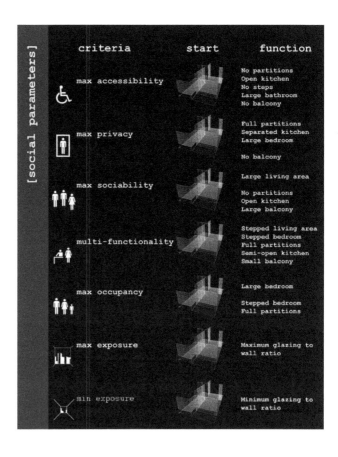

control	input	value min—max	output
partitions' height	> Variable wall_3	zero	
kit. separation relative to livingroom width	> Variable kitchen	zero	
steps height in bedroom and living area	> Variable rise	zero	
bed/bath partition relative to overall depth	> Variable depth_bed	0.4	
balcony depth relative to overall unit depth	> Variable terrace_depth	zero	
partitions' height	> Variables wall_3,wal_2	1.0	
kit. separation relative to livingroom width	> Variable kitchen	0.6	
bedroom/living partitions position relative	> Variable bed_living_n	0.55	
to overall unit width	Variable bed_living_Ot	0.55	
balcony depth relative to overall unit depth	> Variable terrace_depth	zero	
bedroom/living partitions position relative	> Variable bed_living_In	0.3	
to overall unit width	Variable bed_living_Out	0.35	
partitions' height	> Variables wall_3,wall_2	zero	
kit. separation relative to livingroom width	> Variable kitchen	zero	
balcony depth relative to overall unit depth	> Variable terrace_depth	0.2	
steps height in living area	> Variable rise	-1.0	
steps height bedroom	> Variable rise_bed	1.0	
partitions' height	> Variables wall_3,wall_2	1.0	
kit. separation relative to livingroom width	> Variable kitchen	0.5	
balcony depth relative to overall unit depth	> Variable terrace_depth	0.1	
bedroom/living partitions position relative	> Variable bed_living_In	0.4	
to overall unit width	Variable bed_living_Out	0.55	
steps height bedroom	> Variable rise_bed	-1.0	
partitions' height	> Variables wall_3,wall_2	1.0	
horizontal and vertical length of wall rela-	> Variable horizontal	zero	
tive to 1/2 x the unit facade dimensions	Variable vertical	zero	
horizontal and vertical length of wall rela-	> Variable horizontal	0.3	
tive to 1/2 x the unit facade dimensions	Variable vertical	0.4	

3.8 Social system parameters

35

Environmental system

The environmental system primarily controls the position and size of openings of the living unit. Changes are triggered by both the orientation of the unit and its floor location relative to the tower.

Orientation: There is a great potential for this model to accommodate horizontal and vertical sun shading devices. Horizontal glazing divisions eliminate direct sun exposure and minimise the heat gain on south facing facades. Vertical divisions can be parametrically controlled to minimise problematic sun exposure of west facades.

Floor location: Lower floors in high-rise buildings are less exposed to direct and diffused light. Consequently, the lower floors require larger glazing areas. Units on upper levels are normally associated with open views. They, therefore, have unobstructed openings.

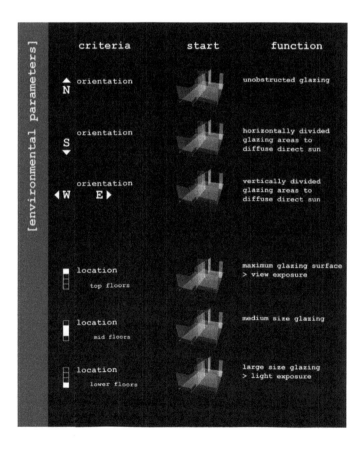

Openings are modeled over the unit's facade at both the bedroom and living spaces. They are controlled parametrically to extend in horizontal and vertical directions and to satisfy the exact form and opening-to-wall ratio. The factors affected are:

1. Horizontal glazing size and spacing
2. Vertical glazing size and spacing

Figure 3.9 describes the environmental influences, functions, and control over the system's glazing components.

control	input	value min——┤——max	output
number of horizontal divisions in the glazing surface	> Variable hz_divisions	1 1├———4	
number of horizontal divisions in the glazing surface	> Variable hz_divisions	4 1├———4	
vertical solid spacing width between glazed panels - in relation to 1/2 x horizontal dimension of each panel	> Variable vertical	0.25 0———0.5	
horizontal / vertical length of solid facade - relative to 1/2 x the unit facade dimensions	> Variable horizontal	zero 0———0.5	
horizontal / vertical length of solid facade - relative to 1/2 x the unit facade dimensions	> Variable horizontal	0.3 0———0.5	
horizontal / vertical length of solid facade - relative to 1/2 x the unit facade dimensions	> Variable horizontal	0.2 0———0.5	

3.9 Environmental system parameters

```
transaction modelBased "Unit points"
{feature In_left GC.Point
    {
    CoordinateSystem            = baseCS;
        XTranslation            = 0;
        YTranslation            = 30;
        ZTranslation            = Series(0,12,12);
        HandlesVisible          = true;
    }
    feature In_right GC.Point
    {
        CoordinateSystem        = baseCS;
        XTranslation            = 20;
        YTranslation            = 30;
        ZTranslation            = Series(0,12,12);
        HandlesVisible          = true;
    }
    feature Out_left GC.Point
    {
        CoordinateSystem        = baseCS;
        XTranslation            = 0;
        YTranslation            = 0;
        ZTranslation            = Series(0,12,12);
        HandlesVisible          = true;
    }
    feature Out_right GC.Point
    {
        CoordinateSystem        = baseCS;
        XTranslation            = 20;
        YTranslation            = 0;
        ZTranslation            = Series(0,12,12);
        HandlesVisible          = true;
    }
}
```

System Making

The following describe the process of creating the parametric design system in *GenerativeComponents*. The unit's components are modeled associated with the rules, constraints, and variables that control the interaction between them.

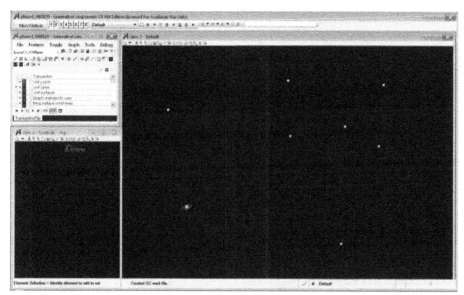

3.10 *Space Configurations* system making 1

Unit points:

Corner points *In_left, In_right, Out_left*, and *Out_right* are created. The dimensions of the unit are 20ft W x 30ft D x 12ft H. Each of the corner points has a floor and ceiling instance: (Z Translation: Series(0,12,12)) with 12 as the height.

```
transaction modelBased "Unit Lines"
{
    feature line01 GC.Line
    {
        StartPoint                 = Out_left;
        EndPoint                   = In_right;
    }
}

transaction modelBased "Unit surfaces"
{
    feature line01 GC.Line
    {
        EndPoint                   = Out_right;
        SymbolXY                   = {103, 103};
    }
    feature line02 GC.Line
    {
        StartPoint                 = In_left;
        EndPoint                   = In_right;
        SymbolXY                   = {100, 103};
    }
    feature line03 GC.Line
    {
        StartPoint                 = In_left;
        EndPoint                   = Out_left;
        SymbolXY                   = {101, 103};
    }
    feature line04 GC.Line
    {
        StartPoint                 = In_right;
        EndPoint                   = Out_right;
        SymbolXY                   = {102, 103};
    }
    feature surface_left GC.BSplineSurface
    {
        Points                     = {In_left,Out_left};
        UcurveDisplay              = 9;
        SymbolXY                   = {99, 102};
    }
    feature surface_right GC.BSplineSurface
    {   Points                     = {In_right,Out_right};
        UcurveDisplay              = 9;
    }
}
```

40

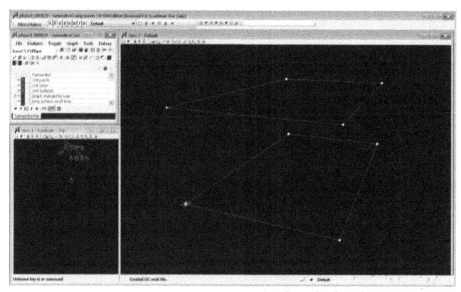

3.11 *Space Configurations* system making 2

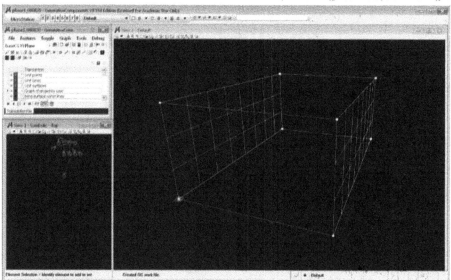

3.12 *Space Configurations* system making 3

Unit surfaces:

Lines between each of the points define the boundary of the model unit. Ruled surfaces *surface_left* and *surface_right* are created on the sides to define the fixed sidewalls.

41

```
transaction modelBased "living surface const lines"
{
    feature In_mid GC.Point
    {
        Curve                   = line02[0];
        T                       = <free> (0.46101271829523);
        SymbolXY                = {98, 105};
        HandlesVisible          = true;
    }
    feature Out_mid GC.Point
    {
        Curve                   = line01[0];
        T                       = <free> (0.437099879921155);
        SymbolXY                = {99, 105};
        HandlesVisible          = true;
    }
    feature bsplineSurface02 GC.BSplineSurface
    {
        StartCurve              = line07;
        EndCurve                = line08;
        UcurveDisplay           = 2;
        VcurveDisplay           = 20;
        SymbolXY                = {97, 108};
    }
}
```

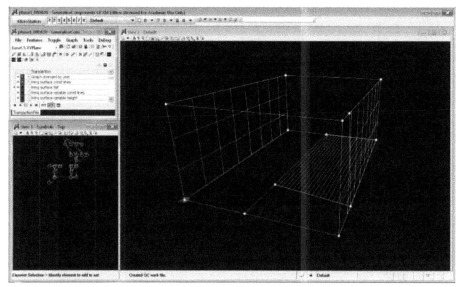

Living surface:

A line on the floor surface divides the unit into 2 zones: a kitchen + living + deck area, and a bedroom + bathroom area. Another line separates the living area from the deck. A surface *Surface02* now defines a flat projection of the living space.

Variable *terrace_depth* controls the deck's size.

Variable *bed_living_In* controls the inner ratio between the two zones.

Variable *bed_living_Out* controls the outer ratio between the two zones.

```
transaction modelBased "living surface variable const lines"
{
    feature In_mid GC.Point
    {
        Curve                   = line02;
    }
    feature Out_mid GC.Point
    {
        Curve                   = line01;
        T                       = <free> (0.50688199162311);
    }
    feature line09 GC.Line
    {
        StartPoint              = point05[0];
        EndPoint                = point05[1];
        SymbolXY                = {98, 108};
    }
    feature line10 GC.Line
    {
        StartPoint              = point07;
        EndPoint                = point06;
        SymbolXY                = {99, 109};
    }
    feature line11 GC.Line
    {
        StartPoint              = In_mid[0];
        EndPoint                = In_mid[1];
        SymbolXY                = {97, 106};
    }
    feature line12 GC.Line
    {
        StartPoint              = In_right[0];
        EndPoint                = In_right[1];
        SymbolXY                = {96, 105};
    }
    feature line13 GC.Line
    {
        StartPoint              = point09;
        EndPoint                = point08;
        SymbolXY                = {96, 110};
    }
    feature point06 GC.Point
    {
        Curve                   = line05;
        T                       = <free> (0.788988246693955);
        SymbolXY                = {101, 109};
        HandlesVisible          = true;
    }
    feature point07 GC.Point
    {
        Curve                   = line09;
        T                       = <free> (0.244748397500457);
        SymbolXY                = {98, 109};
        HandlesVisible          = true;
    }
    feature point08 GC.Point
    {
        Curve                   = line11;
        T                       = <free> (0.22809648460624);
        SymbolXY                = {97, 109};
        HandlesVisible          = true;
    }
    feature point09 GC.Point
    {
        Curve                   = line12;
        T                       = .5058246;
        SymbolXY                = {96, 109};
        HandlesVisible          = true;
    }
}

transaction modelBased "dynamic living surface"
{
    feature bsplineSurface03 GC.BSplineSurface
    {
        StartCurve              = line10;
        EndCurve                = line13;
        UcurveDisplay           = 2;
        VcurveDisplay           = 20;
        SymbolXY                = {97, 112};
    }
    feature line13 GC.Line
    {
        StartPoint              = point08;
        EndPoint                = point09;
    }
}
```

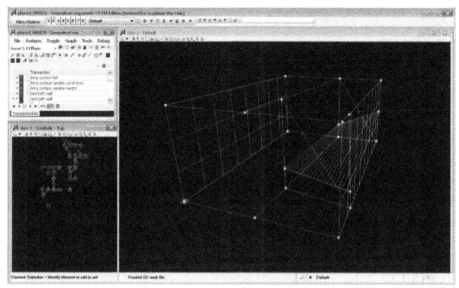

Dynamic living surface:

A number of movable points and guiding lines provide a dynamic base for the living space floor. A floor surface *Surface03* is based on these dynamic points. The points are free to move on vertical guides and, therefore, can change the floor's height at any of the corners.

Bed-Bath transition:

Ruled Surface *Surface04* is created between guiding lines that are free to move along the depth of the unit. It establishes the boundary between the bedroom and bathroom in the unit.

Variable h_In controls the inner floor spot elevation as percentage of full height.

Variable h_Out controls the outer floor spot elevation as percentage of full height.

```
feature rise GC.GraphVariable
    {
        Value                    = -0.26;
        RangeMinimum             = -1.0;
    }
feature stairs GC.GraphVariable
    {
        Value                    = 4.0;
        LimitValueToRange        = true;
        RangeMinimum             = 2.0;
        RangeMaximum             = 5.0;
        RangeStepSize            = 1.0;
        SymbolXY                 = {95, 113};
    }


transaction generateFeatureType "stairs feature created", suppressed
{
    type                    = GC.UBCStairs02;
    inputProperties         = {
                            property GC.CoordinateSystem baseCS
                            {
                                feature              = baseCS;
                                isReplicatable       = true;
                                isParentModel        = true;
                            }
                            property GC.Line line24
                            {
                                feature              = line24;
                                isReplicatable       = true;
                            }
                            property double rise
                            {
                                feature              = rise;
                                isReplicatable       = true;
                            }
                            property double stairs
                            {
                                feature              = stairs;
                                isReplicatable       = true;
                            }
                            };
    outputProperties        = {
                            property GC.Line line18
                            {
                                feature              = line18;
                                isDynamic            = true;
                            }
                            property GC.Line line20
                            {
                                feature              = line20;
                                isDynamic            = true;
                            }
                            property GC.Line line21
                            {
                                feature              = line21;
                                isDynamic            = true;
                            }
                            property GC.Line line22
                            {
                                feature              = line22;
                                isDynamic            = true;
                            }
                            property GC.Point point12
                            {
                                feature              = point12;
                                isConstruction       = true;
                                isDynamic            = true;
                            }
                            };
}
```

46

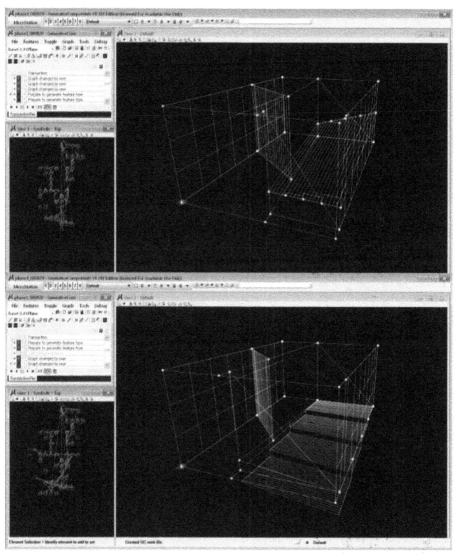

3.15 *Space Configurations* system making 6

Platforms feature:

Platforms are created and applied to the living floor surface based on points, lines, directions, and surfaces.

Variable *rise* controls the height difference between platforms (-1ft to 1ft).

Variable *stairs* controls the number of platforms (2 to5).

```
feature rise_bed GC.GraphVariable
    {
        Value                   = .2;
        LimitValueToRange       = true;
        RangeMinimum            = -1.0;
        RangeMaximum            = 1.0;
        RangeStepSize           = 0.0;
        SymbolXY                = {110, 111};
    }
    feature stairs GC.GraphVariable
    {
        SymbolXY                = {104, 111};
    }
    feature ubcstairs0203 GC.UBCStairs02
    {
        baseCS                  = baseCS;
        line24                  = line29;
        rise                    = rise_bed;
        stairs                  = stairs;
        SymbolXY                = {107, 111};
    }
    feature ubcstairs0204 GC.UBCStairs02
    {
        baseCS                  = baseCS;
        line24                  = line30;
        rise                    = rise_bed;
        stairs                  = stairs;
        SymbolXY                = {106, 111};
    }
}
```

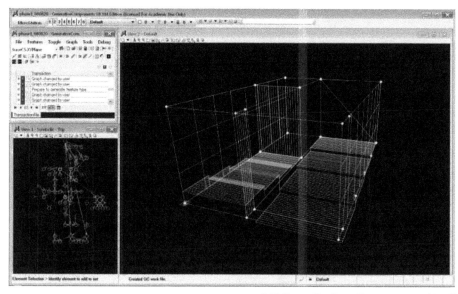

3.16 *Space Configurations* system making 7

Bedroom platforms:

Platforms are constructed over the bedroom floor space. The bedroom entrance (on 1st bedroom platform) has the same height as the 3rd living space platform.

Variable *rise_bed* controls the height difference between bedroom platforms (-1ft to 1ft).

```
transaction modelBased "Bedroom platforms"
{
    feature bsplineSurface11 GC.BSplineSurface
    {
        StartCurve                = ubcstairs0203.line22;
        EndCurve                  = ubcstairs0204.line22;
        UcurveDisplay             = 2;
        VcurveDisplay             = 20;
        SymbolXY                  = {105, 112};
    }
    feature bsplineSurface12 GC.BSplineSurface
    {
        StartCurve                = ubcstairs0203.line21;
        EndCurve                  = ubcstairs0204.line21;
        UcurveDisplay             = 2;
        VcurveDisplay             = 20;
        SymbolXY                  = {108, 112};
    }
    feature bsplineSurface13 GC.BSplineSurface
    {
        StartCurve                = ubcstairs0203.line20;
        EndCurve                  = ubcstairs0204.line20;
        UcurveDisplay             = 2;
        VcurveDisplay             = 20;
        SymbolXY                  = {106, 112};
    }
    feature bsplineSurface14 GC.BSplineSurface
    {
        StartCurve                = ubcstairs0203.line25;
        EndCurve                  = ubcstairs0204.line25;
        UcurveDisplay             = 2;
        VcurveDisplay             = 20;
        SymbolXY                  = {107, 112};
    }
    feature rise_bed GC.GraphVariable
    {
        Value                     = 0.54;
    }
}
```

50

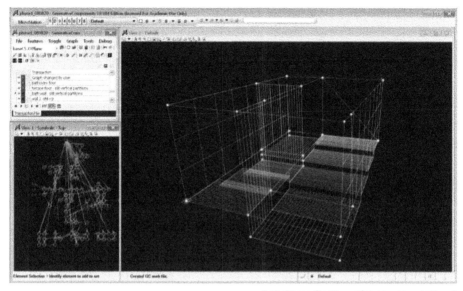

Floors:

The deck shares the same floor height with the 4th platform on the living space, and the bathroom has the same height as the 2nd platform on the living space. These relations depend on the model unit's space layout and the relationships between its components.

```
feature wall_bath GC.GraphVariable
{
    Value                   = 1.0;
    LimitValueToRange       = true;
    RangeMaximum            = 1.0;
    RangeStepSize           = 0.2;
}

feature wall_2 GC.GraphVariable
{
    Value                   = 1.0;
    LimitValueToRange       = true;
    RangeMaximum            = 1.0;
    RangeStepSize           = 0.0;
}
feature wall_bath GC.GraphVariable
{
    Value                   = 0.6;
}

transaction modelBased "vertical ps"
{

    feature wall_3 GC.GraphVariable
    {
        Value                   = 0.62;
        LimitValueToRange       = true;
        RangeMaximum            = 1.0;
        RangeStepSize           = 0.0;
        SymbolXY                = {97, 121};
    }
    feature wall_4 GC.GraphVariable
    {
        Value                   = 0.85;
        LimitValueToRange       = true;
        RangeMaximum            = 1.0;
        RangeStepSize           = 0.0;
        SymbolXY                = {95, 121};
    }
    feature wall_bath GC.GraphVariable
    {
        SymbolXY                = {99, 121};
    }
}
```

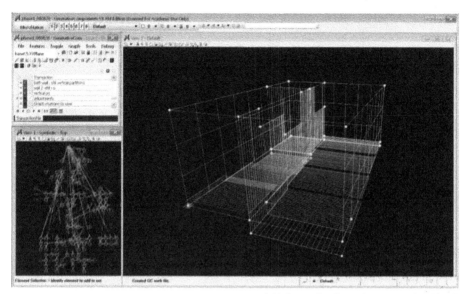

3.18 *Space Configurations* system making 9

Vertical separation:

Partitions are constructed along the floor line that divides the kitchen, living area, and deck from the bedroom and bathroom areas. These are vertically extended from the living space platforms and are each controlled by height variables to a percentage of the full height.

Variable *wall_bath* controls the height of the 1st partition.

Variable *wall_2* controls the height of the 2nd partition.

Variable *wall_3* controls the height of the 3rd partition.

Variable *wall_4* controls the height of the 4th partition.

```
transaction modelBased "skin point grid done"
{
    feature bsplineSurface22 GC.BSplineSurface
    {
        StartCurve              = line47;
        EndCurve                = line26;
        SymbolXY                = {93, 115};
        Visible                 = false;
    }
    feature bsplineSurface23 GC.BSplineSurface
    {
        StartCurve              = line26;
        EndCurve                = line46;
        SymbolXY                = {92, 115};
        Visible                 = false;
    }
        feature point37 GC.Point
        {
        Surface                 = bsplineSurface23;
        U                       = Series(0,1,0.2);
        V                       = Series(0,1,0.2);
        SymbolXY                = {92, 116};
        HandlesVisible          = true;
        Replication             =
ReplicationOption.AllCombinations;
        }
    feature point38 GC.Point
    {
        Surface                 = bsplineSurface22;
        U                       = Series(0,1,0.2);
        V                       = Series(0,1,0.2);
        SymbolXY                = {93, 116};
        Replication             =
ReplicationOption.AllCombinations;
        }
    feature polygon01 GC.Polygon
    {
        Points                  = point37;
        SymbolXY                = {92, 117};
    }
    feature polygon02 GC.Polygon
    {
        Points                  = point38;
        SymbolXY                = {93, 117};
    }
}
```

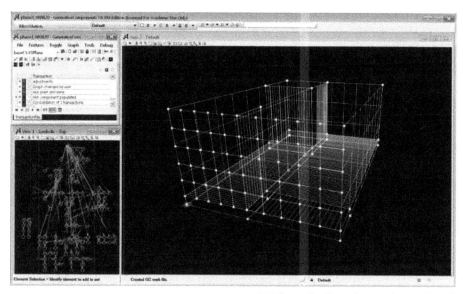

3.19 *Space Configurations* system making 10

Skin point grid:

A surface mesh is constructed over the outer side of the model unit. Points are populated on the surface to form a 5 x 10 grid. The grid size is adjustable.

```
transaction modelBased "skin component populated"
{
    feature horizontal GC.GraphVariable
    {
        Value                      = 0.14;
        LimitValueToRange          = true;
        RangeMaximum               = 0.5;
        RangeStepSize              = 0.0;
        SymbolXY                   = {90, 118};
    }
    feature line48 GC.Line
    {
        StartPoint                 = ubcstairs0201.line22.StartPoint;
        EndPoint                   = point31;
        SymbolXY                   = {99, 120};
    }
    feature point37 GC.Point
    {
        U                          = Series(0,1,0.25);
        V                          = Series(0,1,0.25);
        Visible                    = false;
    }
    feature point38 GC.Point
    {
        U                          = Series(0,1,0.25);
        V                          = Series(0,1,0.25);
        Visible                    = false;
    }
    feature ubcskinComponent0801 GC.UBCSkinComponent08
    {
        horizontal                 = horizontal;
        polygon01                  = polygon01;
        vertical                   = vertical;
        SymbolXY                   = {92, 118};
    }
    feature ubcskinComponent0802 GC.UBCSkinComponent08
    {
        horizontal                 = horizontal;
        polygon01                  = polygon02;
        vertical                   = vertical;
        SymbolXY                   = {93, 118};
        Visible                    = true;
    }
    feature vertical GC.GraphVariable
    {
        Value                      = 0.0;
        LimitValueToRange          = true;
        RangeMaximum               = 0.5;
        RangeStepSize              = 0.0;
        SymbolXY                   = {90, 119};
    }
}
```

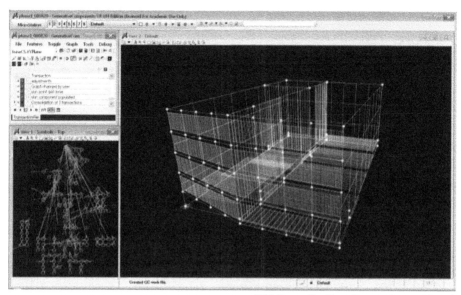

Populated skin component:

To generate skin variations, a model component is created as a surface that occupies each of the grid squares. Variables regulate the horizontal and vertical size of the component. The populated component defines the shape and wall-to-glazing ratio on the unit facade.

Variable *horizontal* controls the horizontal length of surface component.

Variable *vertical* controls the vertical length of surface component.

```
transaction modelBased "terrace side component"
{
    feature point40 GC.Point
    {
        Surface                 = bsplineSurface08;
        U                       = Series(0,1,0.25);
        V                       = Series(0,1,0.25);
        SymbolXY                = {94, 116};
        Replication                                                  =
ReplicationOption.AllCombinations;
        Visible                 = false;
    }
    feature polygon03 GC.Polygon
    {
        Points                  = point40;
        SymbolXY                = {94, 117};
        Visible                 = false;
    }
    feature terrace_depth GC.GraphVariable
    {
        Value                   = .1;
    }
    feature ubcskinComponent0803 GC.UBCSkinComponent08
    {
        horizontal              = horizontal;
        polygon01               = polygon03;
        vertical                = vertical;
        SymbolXY                = {94, 118};
    }
}
```

58

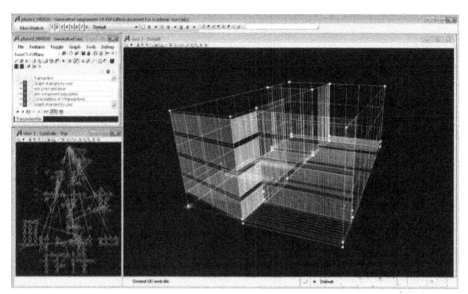

3.21 *Space Configurations* system making 12

To allow for the deck space, a skin surface component is populated on the deck's side. The skin component is now acting on all the exterior walls of the unit.

Variable *horizontal* controls the horizontal length of surface component.

Variable *vertical* controls the vertical length of surface component.

The values of the variables are equal to the glazing length on the side of the skin component (measured as a percentage of the components vertical or horizontal size).

```
feature horizontal GC.GraphVariable
{
    Value           = .1;
}
feature vertical GC.GraphVariable
{
    Value           = 0;
```

59

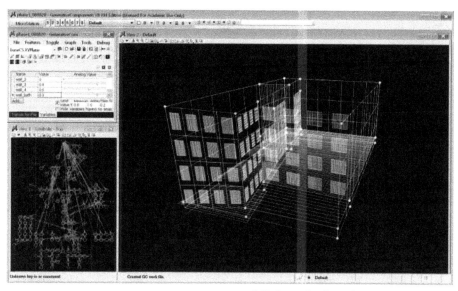

Figure 3.22 shows the effect of manipulating vertical and horizontal lengths on each of the skin glazing components.

```
feature horizontal GC.GraphVariable
{
   Value          = .2;
}
feature vertical GC.GraphVariable
{
   Value          = 0.2;
}
```

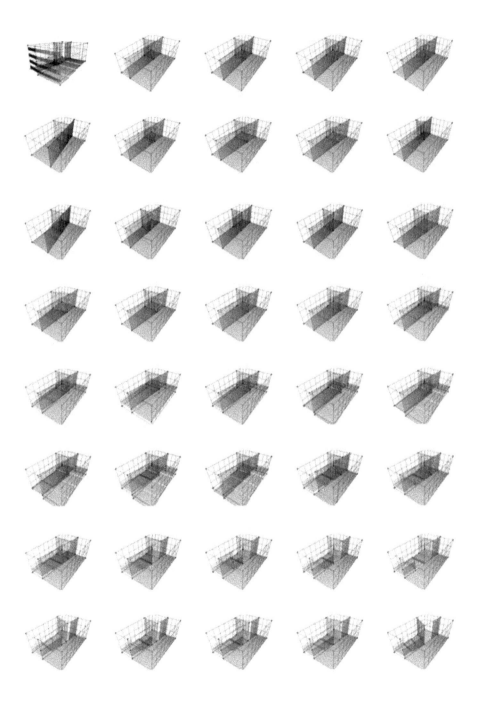

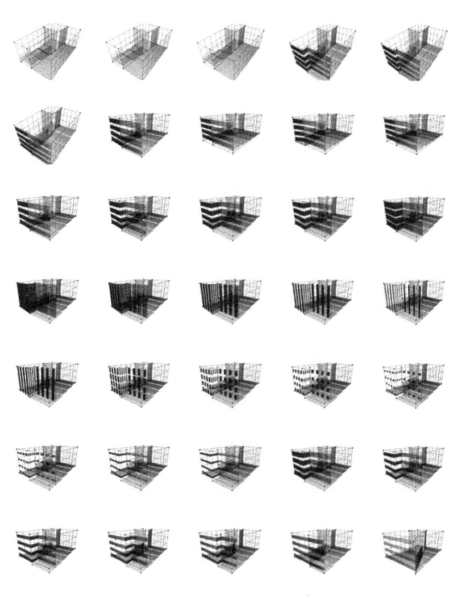

3.23 Instance of model configuration

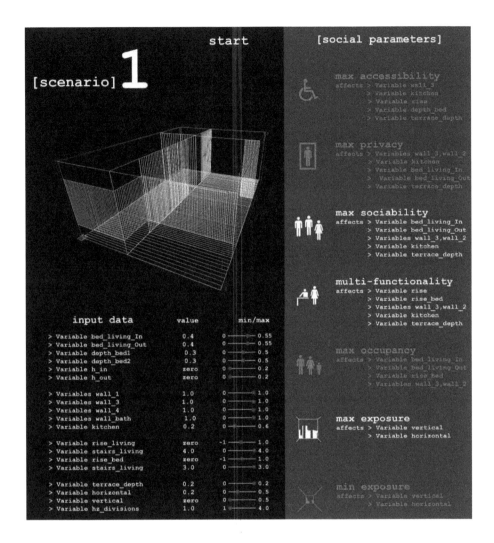

Living Scenarios

The user interface is based on social and environmental factors. Variables are manipulated to find the most suitable configuration for the desired factors. Figures 3.24, 2.25, and 3.26 describe potential living scenarios that combine different social and environmental criteria. The illustrations show the values for variables that reconfigure the system's spatial components according to the established rules. The design system responds to the change of parameters and dynamically reconfigures its components to find a non-linear accumulation of spatial results.

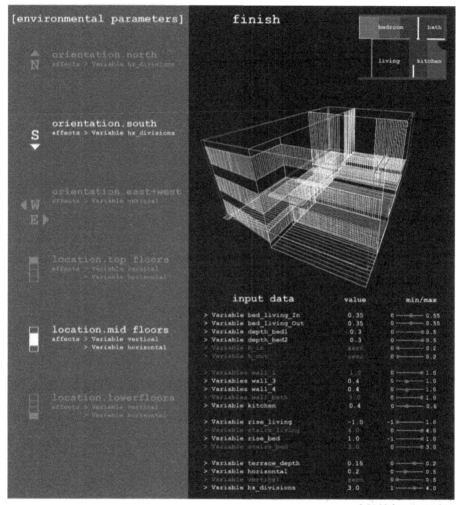

3.24 Living scenario 1

For example, in scenario 1 (figure 3.24), occupants choose 'Maximum Sociability', 'Multi-Functionality', and 'Maximum Exposure' as references for the unit design. The unit is south facing and located in the middle floors of the tower. The maximum sociability parameter corresponds to the size of the living area, spaces are stepped to facilitate multi-functionality, and the glazing is horizontally divided while maintaining a large area to accommodate for southern orientation and maximum exposure.

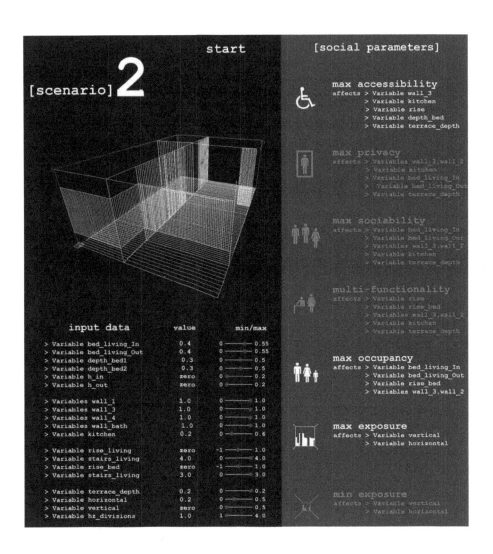

start

[scenario] 2

[social parameters]

max accessibility
affects > Variable wall_3
> Variable kitchen
> Variable rise
> Variable depth_bed
> Variable terrace_depth

max privacy
affects > Variables wall_3,wall_2
> Variable kitchen
> Variable bed_living_In
> Variable bed_living_Out
> Variable terrace_depth

max sociability
affects > Variable bed_living_In
> Variable bed_living_Out
> Variables wall_3,wall_2
> Variable kitchen
> Variable terrace_depth

multi-functionality
affects > Variable rise
> Variable rise_bed
> Variables wall_3,wall_2
> Variable kitchen
> Variable terrace_depth

max occupancy
affects > Variable bed_living_In
> Variable bed_living_Out
> Variable rise_bed
> Variables wall_3,wall_2

max exposure
affects > Variable vertical
> Variable horizontal

min exposure
affects > Variable vertical
> Variable horizontal

input data	value	min/max	
> Variable bed_living_In	0.4	0	0.55
> Variable bed_living_Out	0.4	0	0.55
> Variable depth_bed1	0.3	0	0.5
> Variable depth_bed2	0.3	0	0.5
> Variable h_in	zero	0	0.2
> Variable h_out	zero	0	0.2
> Variables wall_1	1.0	0	1.0
> Variables wall_3	1.0	0	1.0
> Variables wall_4	1.0	0	1.0
> Variables wall_bath	1.0	0	1.0
> Variable kitchen	0.2	0	0.6
> Variable rise_living	zero	-1	1.0
> Variable stairs_living	4.0	0	4.0
> Variable rise_bed	zero	-1	1.0
> Variable stairs_living	3.0	0	3.0
> Variable terrace_depth	0.2	0	0.2
> Variable horizontal	0.2	0	0.5
> Variable vertical	zero	0	0.5
> Variable hz_divisions	1.0	1	4.0

66

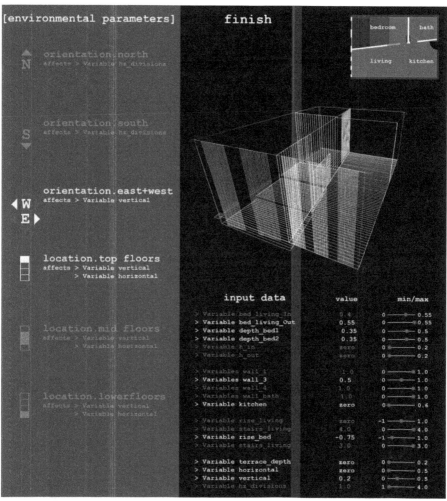

[environmental parameters]   finish

bedroom   bath

living   kitchen

N orientation.north
   affects > Variable hz_divisions

S orientation.south
   affects > Variable hz_divisions

◀ W   orientation.east+west
  E ▶ affects > Variable vertical

location.top floors
affects > Variable vertical
        > Variable horizontal

location.mid floors
affects > Variable vertical
        > Variable horizontal

location.lowerfloors
affects > Variable vertical
        > Variable horizontal

input data	value	min/max
> Variable bed_living_In	0.4	0 —— 0.55
> Variable bed_living_Out	0.55	0 —— 0.55
> Variable depth_bed1	0.35	0 —— 0.5
> Variable depth_bed2	0.35	0 —— 0.5
> Variable h_in	zero	0 —— 0.2
> Variable h_out	zero	0 —— 0.2
> Variables wall_1	1.0	0 —— 1.0
> Variables wall_3	0.5	0 —— 1.0
> Variables wall_4	1.0	0 —— 1.0
> Variables wall_bath	1.0	0 —— 1.0
> Variable kitchen	zero	0 —— 0.6
> Variable rise_living	zero	-1 —— 1.0
> Variable stairs_living	4.0	0 —— 4.0
> Variable rise_bed	-0.75	-1 —— 1.0
> Variable stairs_living	3.0	0 —— 3.0
> Variable terrace_depth	zero	0 —— 0.2
> Variable horizontal	zero	0 —— 0.5
> Variable vertical	0.2	0 —— 0.5
> Variable hz_divisions	1.0	1 —— 4.0

3.25 Living scenario 2

67

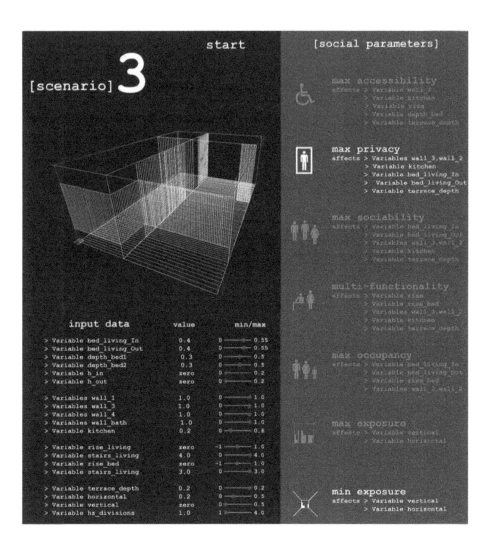

start

[scenario] **3**

[social parameters]

max accessibility
affects > Variable wall_3
> Variable Kitchen
> Variable rise
> Variable depth_bed
> Variable terrace_depth

max privacy
affects > Variables wall_3,wall_2
> Variable kitchen
> Variable bed_living_In
> Variable bed_living_Out
> Variable terrace_depth

max sociability
affects > Variable bed_living_In
> Variable bed_living_Out
> Variables wall_3,wall_2
> Variable kitchen
> Variable terrace_depth

multi-functionality
affects > Variable rise
> Variable rise_bed
> Variables wall_3,wall_2
> Variable kitchen
> Variable terrace_depth

max occupancy
affects > Variable bed_living_In
> Variable bed_living_Out
> Variable rise_bed
> Variables wall_3,wall_2

max exposure
affects > Variable vertical
> Variable horizontal

min exposure
affects > Variable vertical
> Variable horizontal

input data	value	min/max
> Variable bed_living_In	0.4	0 — 0.55
> Variable bed_living_Out	0.4	0 — 0.55
> Variable depth_bed1	0.3	0 — 0.5
> Variable depth_bed2	0.3	0 — 0.5
> Variable h_in	zero	0 — 0.2
> Variable h_out	zero	0 — 0.2
> Variables wall_1	1.0	0 — 1.0
> Variables wall_3	1.0	0 — 1.0
> Variables wall_4	1.0	0 — 1.0
> Variables wall_bath	1.0	0 — 1.0
> Variable kitchen	0.2	0 — 0.6
> Variable rise_living	zero	-1 — 1.0
> Variable stairs_living	4.0	0 — 4.0
> Variable rise_bed	zero	-1 — 1.0
> Variable stairs_living	3.0	0 — 3.0
> Variable terrace_depth	0.2	0 — 0.2
> Variable horizontal	0.2	0 — 0.5
> Variable vertical	zero	0 — 0.5
> Variable hz_divisions	1.0	1 — 4.0

68

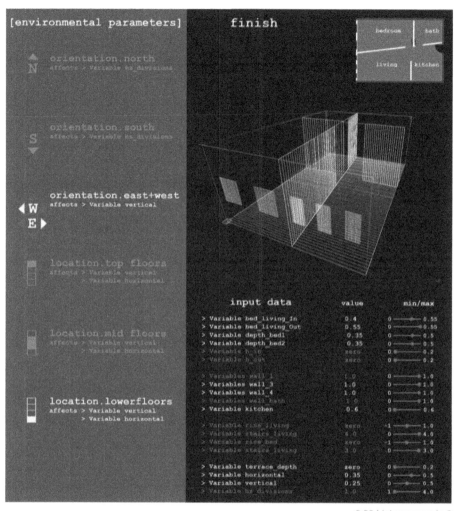

[environmental parameters]

finish

| | bedroom | bath |
| living | kitchen |

▲
N orientation.north
affects > Variable hs_divisions

S orientation.south
affects > Variable hs_divisions
▼

orientation.east+west
◀ W affects > Variable vertical
E ▶

location.top floors
affects > Variable vertical
> Variable horizontal

location.mid floors
affects > Variable vertical
> Variable horizontal

location.lowerfloors
affects > Variable vertical
> Variable horizontal

input data	value	min/max
> Variable bed_living_In	0.4	0 — 0.55
> Variable bed_living_Out	0.55	0 — 0.55
> Variable depth_bed1	0.35	0 — 0.5
> Variable depth_bed2	0.35	0 — 0.5
> Variable h_in	zero	0 — 0.2
> Variable h_out	zero	0 — 0.2
> Variables wall_1	1.0	0 — 1.0
> Variables wall_3	1.0	0 — 1.0
> Variables wall_4	1.0	0 — 1.0
> Variables wall_bath	1.0	0 — 1.0
> Variable kitchen	0.6	0 — 0.6
> Variable rise_living	zero	-1 — 1.0
> Variable stairs_living	4.0	0 — 4.0
> Variable rise_bed	zero	-1 — 1.0
> Variable stairs_living	3.0	0 — 3.0
> Variable terrace_depth	zero	0 — 0.2
> Variable horizontal	0.35	0 — 0.5
> Variable vertical	0.25	0 — 0.5
> Variable hs_divisions	1.0	1 — 4.0

3.26 Living scenario 3

69

Comments

The explorations for *Space Configurations* illustrate the potential to integrate procedural thinking into the design process. The parametric design system in this study is responsive to various functional concerns and parameters related to dense residential spaces. Components of space correspond to variables that can be manually controlled to initiate a reconfiguration of the overall system.

The design system supports a wide range of spatial configurations. These spatial variations constitute an accumulated progression of the design system's preset system components. Due to the responsiveness of the design system, after-the fact evaluation of the produced configurations is not necessary. The parametric design approach only generates feasible space configurations because design limitations are incorporated in the rules for the design system.

The initial decision to address functional, programmatic concerns, usability, and accessibility is reflected in the array of spatial configuration explored with the help of the design system. The developed parametric design system can be further utilised in the design of a residential high-rise where the configuration of customized units affect the design of the entire structure.

The digital techniques used in the design are set to incorporate design constraints into the process of design. This informs the process of design by directing the output towards the designer's intended direction. This approach generates responsive architecture that is capable of addressing a magnitude of design concerns. An array of individual design decisions result in a complex design process governed by interrelated parameters.

A user interface directly linked to parameters of a *GenerativeComponents* parametric design system can allow potential occupants to choose infinite combinations of occupation and location specific to their needs and review the resulting spaces.

[iv] City of Vancouver, Planning Department, Statistics and Information
http://vancouver.ca/commsvcs/planning/stats.htm
Screen clipping taken: 28/10/2008, 2:38 PM

70

4. Tower Formation

Introduction

This exploration makes use of computation and the logic of emergence. In order to illustrate this approach to design, the exploration works on the scale of a building in an urban setting. The goal is to generate formal complexity out of a simple initial design, while dynamically responding to a variety of forces that influence the overall form of a building. Similar to the first design exploration, the design methodology to *Tower Formation* relies on rule-based procedures.

Design Approach

Instead of considering each design constraint separately, and working through a regular design process, the parametric model used in this exploration allows responding to all parameters simultaneously. As a result, the design is responsive to a complex set of influences. The designer has control over how the parametric model reacts to each parameter by setting rules and can review the interaction of the influencing factors in the digital model. This computational modelling process permits engaging with layers of design complexity that cannot be incorporated into conventional CAD models.

The focus of this exploration is the design of a residential tower using parametric design techniques. The form of the tower is directly influenced by contextual parameters. Context, orientation and views influence a dynamic design system to generate the building's configuration. As soon as the rules of how the system responds to each of the parameters are defined, the system calculates and graphically represents the influences, and generates alterations to its form.

In principle, the proposed design system is not limited to the relatively small number of influencing parameters considered in this exploration of a high-rise residential design.

Design Background

This exploration focuses on the design of a residential tower in Vancouver's downtown core. While residential towers have been successful in increasing density and meeting the city's planning goals, tower designs do not engage in contemporary design concepts and design and construction techniques. As a result, the existing towers have remained generic applications of the tower typology without engaging in the particularities of physical surroundings.

The design approach proposed in this study provides a conceptual alternative to the design of existing high-rise buildings. Context parameters govern the design process and contribute to the formation of the tower design.

Objective

The objective of this study is to create a dynamic responsive system able to interact with the immediate environment of the building. A parametric design system controls the building's form. The tower form is adjusted according to changes of the influencing factors. This approach allows for designs that are responsive to an urban context that is constantly changing.

The success of the design approach is tested through the responsiveness of the design to its urban context.

Site

The site for this exploration is located in downtown Vancouver. The existing site includes three recently constructed towers and a public park.

The site has a number of characteristics typical for a downtown Vancouver location:

1. Adjacency to various high-rise and low-rise buildings.

2. Views to significant parts of the city - The exploration include studies on how views from the residential units can affect the form of the building.

3. View Corridor 9.2.1 - The site is crossed by one of the view corridors assigned by the City of Vancouver as a reserved view. Building developments are not allowed to block reserved view corridors. Regulations for building in view corridors are incorporated in the formation process of the tower.

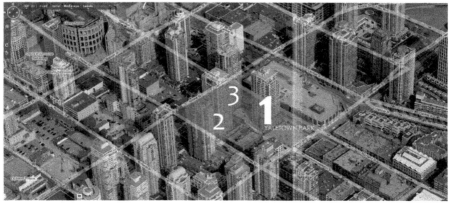

4. Park - The site for the tower includes a park. The relationship between the park and tower is explored.

4.1 Site map

73

Existing Architecture

The existing tower at the site, designed by Buttjes Architecture, is typical of downtown Vancouver. It is 30-story tower that primarily contains small one-bedroom units and townhouses in the tower podium. The design with its amenities, aesthetic approach, and efficient use of spaces meets current city planning requirements and market demands.

The typical Vancouver residential tower design does not respond to contextual conditions. The living units are designed for maximum exposure regardless of their location within the building. The building mass can be easily reduced to an extruded square with minimal response to the immediate and extended context.

4.2 Existing architecture on site

Position

Tower Formation is based on a design approach that focuses on a dynamic form-finding process. Program, site, and context are parameters for the design. The form-finding process takes three primary factors into account:

1. View corridor

2. Orientation

3. Context

Methodology

Parametric modeling in *GenerativeComponents* is used as a platform to explore complex relationships of context parameters. The building form results from the negotiation of these factors. With every additional parameter, the design dramatically increases in complexity.

74

Tower Formation takes the existing tower volume as a starting point. The proposed form is based on the original building volume that achieves a high degree of density. The integration of context parameters into the design process results in deformations of the building volume. This approach ultimately leads to a building configuration that is responsive to context parameters. Floor plan layouts are subsequently affected by the change in building form.

Design Process

The high-rise tower is designed as a dynamic model in *GenerativeComponents*. The initial building volume is determined by considering existing setback and height regulations sun exposure, and crossing view corridors. The system is set to respond to changes of these parameters in real time, thus allowing for experimentation at model scale.

For each of the parameters, rules set inside the software regulate the response of the building.

Initial Form: The initial form of the tower is the most basic extruded rectangle that results from considering setback regulations and the allowable building height.

Factor 1: View Corridors

In 1991, the city of Vancouver assigned a number of view corridors to preserve views to the North Shore Mountains from a number of locations in the city[v]. Building heights and forms are not to obstruct the view corridors.

4.3 Site context

The proposed design system reacts to the view cone, not only by avoiding obstruction, but also by engaging the building with the preserved mountains view.

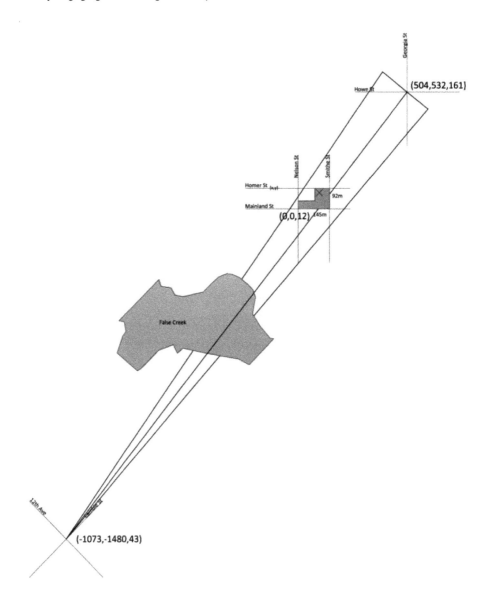

Factor 2: Urban Context

The site is surrounded by a number of high-rise and low-rise buildings. The spaces between the surrounding buildings offer open views to different parts of the city. Two prospective in-between views from the tower are considered to configure the tower.

In order to engage with these views, a V-shaped cut provides additional façade surfaces oriented towards the in-between spaces and views. The width of the cut is proportional to the dimension of the in-between spaces. The cut is deeper at the top of the tower to provide more exposure towards the top of the tower.

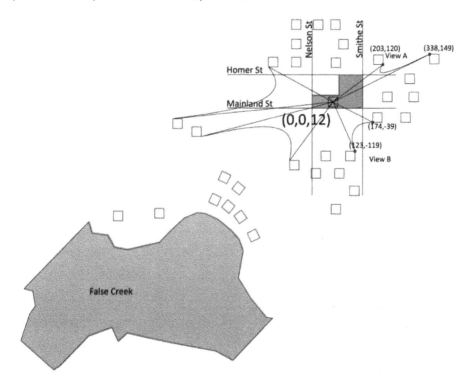

Factor 3: Orientation

The tower is designed to maximise south exposure. It is configured for maximum passive solar heating by maximising southern exposures while maintaining a relatively compact building volume.

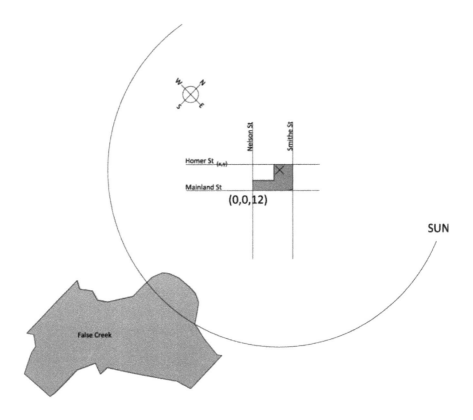

Factor 4: Context at Skin Scale

The tower is configured to allow for open views at the top of the building and to provide for less exposure immediately adjacent to other buildings. At the same time, floor levels near the base of the tower are configured for increased light exposures. Facade components are adjusted to accommodate changing light conditions.

Tower configuration, floor heights, and façade treatment contribute collectively to form finding process. Each parameter influences the form of the building by contributing to a dynamic shaping process.

Figure 4.7 describes the formation process of the tower. The diagrams show how the form of the tower responds to each of the individual influencing criteria (parameters).

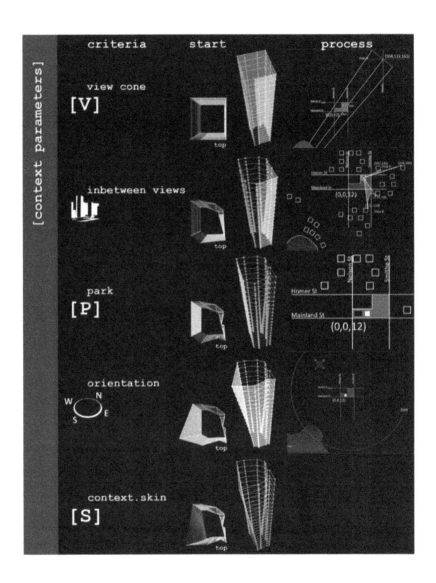

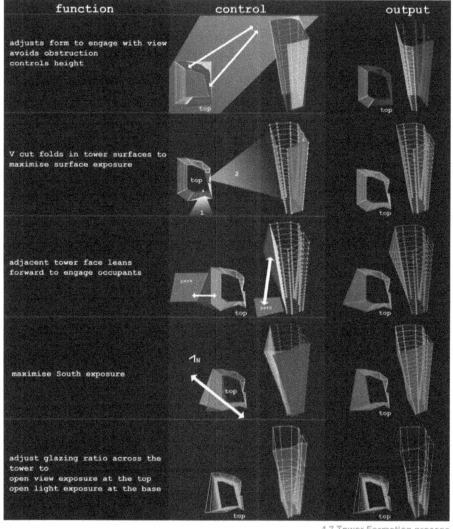

function	control	output
adjusts form to engage with view avoids obstruction controls height		
V cut folds in tower surfaces to maximise surface exposure		
adjacent tower face leans forward to engage occupants		
maximise South exposure		
adjust glazing ratio across the tower to open view exposure at the top open light exposure at the base		

4.7 Tower Formation process

81

The following describe the process of creating the parametric design system in *GenerativeComponents*. The model is controlled by rules, constraints, and variables that define the interaction between components.

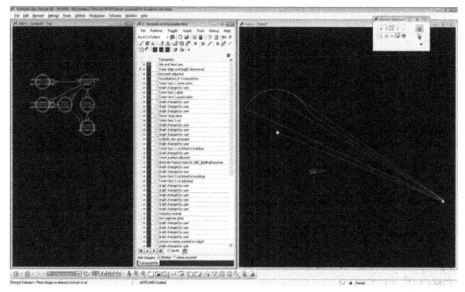

4.8 *Tower Formation* system making 1

Site Specifications:

Site starting point: Vancouver, Nelson Street at Mainland Street

Site dimensions: 145 m x 92 m

Site elevation: 12 m average

View corridor 9.2.1 is drawn in relation to the site location.

View corridor reference data:

Preserved view: North Shore Mountains

View Point: Cambie Street at 12th Avenue (43.47 m elevation)

Reference Point 9.2.1: T.D. Bank Tower (161.73 m elevation)

Distance from view of point to reference point: 2419.80 m

```
transaction modelBased "Tower edge and height determined"
{
    feature Setback GC.GraphVariable
    {
        Value                    = 3.0;
        LimitValueToRange        = true;
        RangeMaximum             = 10.0;
        RangeStepSize            = 1.0;
        SymbolXY                 = {99, 103};
    }
    feature floorheight GC.GraphVariable
    {
        Value                    = 3;
        LimitValueToRange        = true;
        RangeMinimum             = 2.5;
        RangeMaximum             = 5.0;
        RangeStepSize            = 0.5;
        SymbolXY                 = {99, 104};
    }
    feature p1 GC.Point
    {
        CoordinateSystem         = baseCS;
        Xtranslation             = point04.X+Setback;
        Ytranslation             = point04.Y-Setback;
        Ztranslation             =
Series(point04.Z,point06.Z,floorheight);
        SymbolXY                 = {101, 104};
        HandleDisplay            = DisplayOption.Display;
    }
    feature point03 GC.Point
    {
        CoordinateSystem         = baseCS;
        Xtranslation             = 0;
        Ytranslation             = 0;
        Ztranslation             = 12;
        SymbolXY                 = {99, 100};
        HandleDisplay            = DisplayOption.Display;
    }
    feature point04 GC.Point
    {
        CoordinateSystem         = baseCS;
        Xtranslation             =
uBCTowerSitePLan01.line04.StartPoint.X;
        Ytranslation             =
uBCTowerSitePLan01.line04.StartPoint.Y;
        Ztranslation             =
uBCTowerSitePLan01.line04.StartPoint.Z;
        SymbolXY                 = {101, 100};
        HandleDisplay            = DisplayOption.Display;
    }
    feature point06 GC.Point
    {
        Curve                    = uBCViewCone01.View;
        PointToProjectOntoCurve  = point04[0];
        SymbolXY                 = {100, 103};
    }
    feature uBCTowerSitePLan01 GC.UBCTowerSitePLan
    {
        SiteOrigin               = point03;
    }
}

transaction modelBased "first point adjusted"
{
    feature point04 GC.Point
    {
        Xtranslation             =
uBCTowerSitePLan01.line06.StartPoint.X;
        Ytranslation             =
uBCTowerSitePLan01.line06.StartPoint.Y;
        Ztranslation             =
uBCTowerSitePLan01.line06.StartPoint.Z;
    }
}

transaction modelBased "Consolidation of 3 transactions"
{
    feature Parkdepth GC.GraphVariable
    {
        Value                    = 35;
        LimitValueToRange        = true;
        RangeMaximum             = 40.0;
        RangeStepSize            = 0.0;
    }
    feature p1 GC.Point
    {
        Xtranslation             = point04.X+Setback+Parkdepth;
    }
```

84

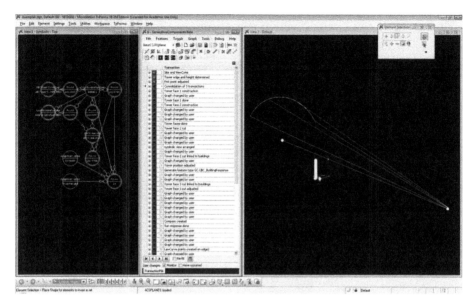

4.9 *Tower Formation* system making 2

First Edge:

The first edge of the tower *p1* is placed within a variable distance setback from the corner of the site.

The height of the building *height* is a value determined by the difference in elevation between:

a. The Site's projection on the view corridor, and

b. The site's plane at 12 m

Floor heights are set as a global variable *floor height*.
When *floor height* is 3 meters, the calculated number of floors equals 40.

```
transaction modelBased "Tower face 1 construction"
{
    feature Setback GC.GraphVariable
    {
        RangeMaximum              = 1.0;
    }
    feature Tview GC.GraphVariable
    {
        Value                     = 0.76;
        LimitValueToRange         = true;
        RangeMaximum              = 1.0;
        RangeStepSize             = 0.0;
        SymbolXY                  = {99, 106};
    }
    feature View GC.GraphVariable
    {
        Value                     = 26.0;
        LimitValueToRange         = true;
        RangeStepSize             = 0.0;
        SymbolXY                  = {99, 105};
    }
    feature bsplineCurve01 GC.BsplineCurve
    {
        FitPoints                 = {p1};
        SymbolXY                  = {102, 104};
    }
    feature line01 GC.Line
    {
        StartPoint                = p1;
        EndPoint                  = line02.EndPoint;
        SymbolXY                  = {100, 106};
    }
    feature line02 GC.Line
    {
        StartPoint                = pview;
        Direction                 = baseCS.Zdirection;
        Length                    = -View;
        SymbolXY                  = {100, 104};
    }
    feature line03 GC.Line
    {
        StartPoint                = point09;
        EndPoint                  = p1a;
        SymbolXY                  = {102, 107};
    }
    feature p1a GC.Point
    {
        CoordinateSystem          = baseCS;
        Xtranslation              = p1.X+5;
        Ytranslation              = p1.Y;
        Ztranslation              = p1.Z;
        SymbolXY                  = {101, 106};
        HandleDisplay             = DisplayOption.Display;
    }
    feature p1b GC.Point
    {
        CoordinateSystem          = baseCS;
        Xtranslation              = p1.X+20;
        Ytranslation              = p1.Y-line03.Length;
        Ztranslation              = p1.Z;
        SymbolXY                  = {102, 106};
    }
    feature point09 GC.Point
    {
        Curve                     = line01;
        PointToProjectOntoCurve   = p1a;
        SymbolXY                  = {102, 108};
    }
    feature pview GC.Point
    {
        Curve                     = uBCViewCone01.View;
        T                         = Tview;
        SymbolXY                  = {100, 107};
        HandleDisplay             = DisplayOption.Display;
    }
    feature side GC.GraphVariable
    {
        Value                     = 18;
        LimitValueToRange         = true;
        RangeMinimum              = 10.0;
        RangeMaximum              = 30.0;
        RangeStepSize             = 1.0;
        SymbolXY                  = {99, 107};
    }
}
```

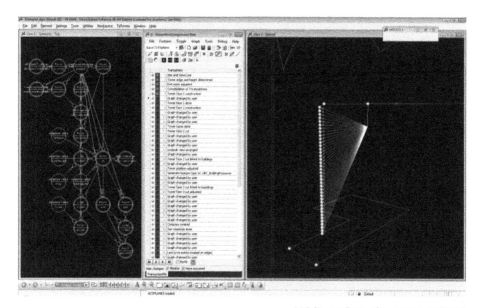

4.10 *Tower Formation* system making 3

The tower's edge point series *p1* is connected by a line *line01* to the view corridor's reference point (target).

Tview and *View* are variables connected to the target point of the view cone. Their manipulation directs the orientation of the tower towards the view target.

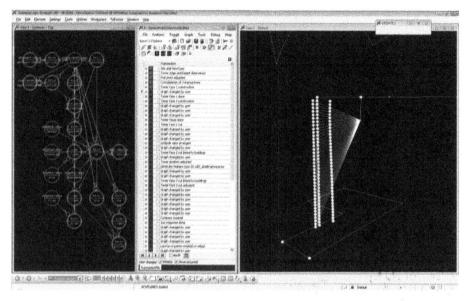

View cone response rule:

The tower's faces adjust to better engage with the view. The points defining each of the tower's faces move to provide greater surface exposure to the view target.

Each surface is defined by 4 series of points:

2 edges + 2 isocurves on the face

The second isocurve's *p1b* position moves along the y-axis to direct the building face towards the view target. The distance moved equals the distance between the first isocurve *p1a* and *line01*. (Figure 4.11)

```
transaction modelBased "Graph changed by user"
{
    feature View GC.GraphVariable
    {
        Value                        = 29;
    }
}

transaction modelBased "Graph changed by user"
{
    feature line04 GC.Line
    {
        StartPoint                   = point11;
        EndPoint                     = pla;
    }
    feature line04_EndPoint GC.Point
    {
        Plane                        = baseCS.XYplane;
        Xtranslation                 = <free> (507.950808438218);
        Ytranslation                 = <free> (460.042810956901);
    }
    feature p2 GC.Point
    {
        CoordinateSystem             = baseCS;
        Xtranslation                 = p1.X+30;
        Ytranslation                 = p1.Y-line04.Length;
        Ztranslation                 = p1.Z;
        HandleDisplay                = DisplayOption.Display;
    }
    feature point11 GC.Point
    {
        Curve                        = line01;
        PointToProjectOntoCurve      = plb;
    }
}

transaction modelBased "Tower face 1 done"
{
    deleteFeature line04_EndPoint;
    feature Tview GC.GraphVariable
    {
        Value                        = 0.7599;
        RangeMinimum                 = 0.75;
        RangeMaximum                 = 0.78;
    }
    feature bsplineCurve01 GC.BsplineCurve
    {
        SymbolXY                     = {103, 104};
    }
    feature c1a GC.BsplineCurve
    {
        FitPoints                    = {pla};
        SymbolXY                     = {102, 109};
    }
    feature point09 GC.Point
    {
        SymbolXY                     = {101, 107};
    }
}
```

90

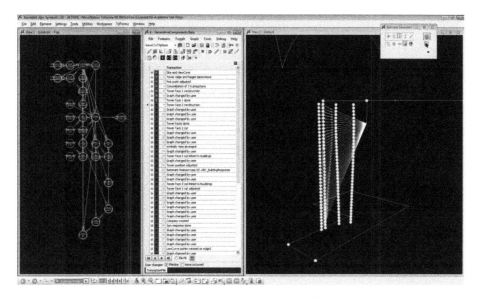

4.12 *Tower Formation* system making 5

The second edge of the tower *p2* is equally adjusted along the y-axis. The distance moved equals the distance between the second isocurve *p1b* and *line01*. The variable *Side* determines the width of the building surface. Spacing between each of the surface curves is a factor of this variable.

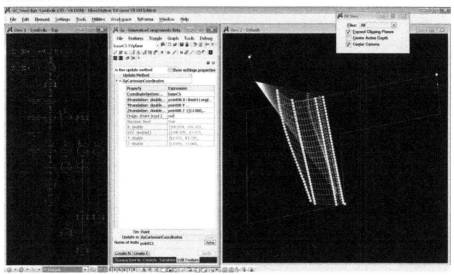

4.13 *Tower Formation* system making 6

A surface mesh *s12* lofts the curves to define the first face of the tower.

91

```
transaction modelBased "Tower face 2 construction"
{
    feature p1a GC.Point
    {
        Xtranslation              = p1.X+side*.25;
    }
    feature p1b GC.Point
    {
        Xtranslation              = p1a.X+side*.45;
    }
    feature p2 GC.Point
    {
        Xtranslation              = p1b.X+line04.Length*2-side*.6;
        Ytranslation              = p1b.Y-line04.Length*2+side*.65;
    }
    feature p3 GC.Point
    {
        CoordinateSystem          = baseCS;
        Xtranslation              = p1.X+side;
        Ytranslation              = p1.Y-side;
        Ztranslation              = p1.Ztranslation;
    }
    feature point13 GC.Point
    {
        CoordinateSystem          = baseCS;
        Xtranslation              = p1.X+side;
        Ytranslation              = p1.Y-side*.75;
        Ztranslation              = p1.Ztranslation;
    }
    feature side GC.GraphVariable
    {
        Value                     = 20;
    }
}

transaction modelBased "Graph changed by user"
{
    feature bsplineCurve12 GC.BsplineCurve
    {
        FitPoints                 = {p1b};
    }
    feature bsplineCurve14 GC.BsplineCurve
    {
        FitPoints                 = {p3};
        SymbolXY                  = {104, 110};
    }
    feature c1a GC.BsplineCurve
    {
        SymbolXY                  = {101, 108};
    }
    feature c2 GC.BsplineCurve
    {
        FitPoints                 = {p2};
        SymbolXY                  = {98, 110};
    }


transaction modelBased "Graph changed by user"
{
    feature bsplineSurface02 GC.BsplineSurface
    {
        Curves                                                              =
{bsplineCurve01[0],c1a[0],bsplineCurve12[0],c2[0]};
        Order                     = 3;
        UcurveDisplay             = 20;
        VcurveDisplay             = 20;
    }
}
```

92

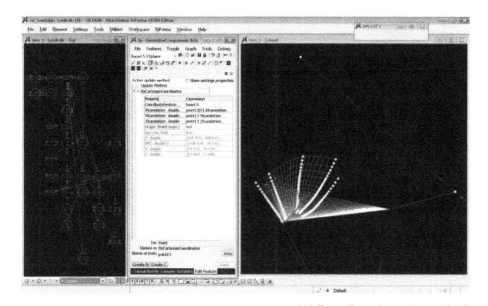

4.14 *Tower Formation* system making 7

4.15 *Tower Formation* system making 8

P2 and *P3* (second and third corner point series) are each located at a distance equal to the side of the building *side* from *P1* (first corner point series).The isocurve along the second face is determined by its relative distance between *P2* and *P3*.

```
transaction modelBased "Graph changed by user"
{
    feature bsplineSurface01 GC.BsplineSurface
    {
        Curves              = {c3[0],bsplineCurve14[0]};
        Order               = 3;
        UcurveDisplay       = 20;
        VcurveDisplay       = 20;
        SymbolXY            = {104, 111};
    }
    feature bsplineSurface04 GC.BsplineSurface
    {
        Curves              = {bsplineCurve01[0],c4[0]};
        UcurveDisplay       = 20;
        VcurveDisplay       = 20;
        SymbolXY            = {102, 111};
    }
    feature c3 GC.BsplineCurve
    {
        FitPoints           = {p3};
        SymbolXY            = {101, 108};
    }
    feature c4 GC.BsplineCurve
    {
        FitPoints           = {p4};
        SymbolXY            = {99, 110};
    }
    feature p4 GC.Point
    {
        CoordinateSystem    = baseCS;
        Xtranslation        = p1.X;
        Ytranslation        = p1.Y-side;
        Ztranslation        = p1.Z;
        SymbolXY            = {99, 109};
    }
    feature s_cd GC.BsplineSurface
    {
        Curves              = {c3[0],c4[0]};
        Order               = 3;
        UcurveDisplay       = 20;
        VcurveDisplay       = 20;
        SymbolXY            = {101, 111};
    }
}

transaction modelBased "Tower faces done"
{
    deleteFeature bsplineCurve01;
    deleteFeature bsplineSurface04;

    feature bsplineCurve08 GC.BsplineCurve
    {
        FitPoints           = {p4};
    }
    feature bsplineSurface01 GC.BsplineSurface
    {
        Curves              = {c3[0],c4[0]};
    }
    feature c1 GC.BsplineCurve
    {
        FitPoints           = {p1};
    }
    feature c1b GC.BsplineCurve
    {
        FitPoints           = {p1b};
    }
    feature c222222 GC.BsplineCurve
    {
        FitPoints           = {p2};
    }
    feature s_ab GC.BsplineSurface
    {
        Curves              = {c1[0],c1a[0],c1b[0],c222222[0]};
        UcurveDisplay       = 20;
        VcurveDisplay       = 20;
    }
    feature s_bc GC.BsplineSurface
    {
        Curves              = {c222222[0],c3[0]};
        UcurveDisplay       = 20;
        VcurveDisplay       = 20;
    }
    feature s_da GC.BsplineSurface
    {
        Curves              = {c1[0],c4[0]};
        UcurveDisplay       = 20;
        VcurveDisplay       = 20;
    }
}
```

94

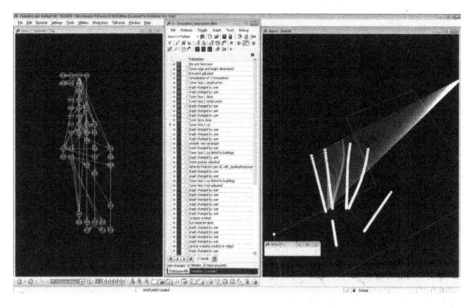

4.16 *Tower Formation* system making 9

The third corner point series *p3* is positioned (side,-side) relative to the first corner *p1*.
The fourth corner point series *p4* is positioned (0,-side) relative to the first corner *p1*.
Surfaces *(s_ab, s_bc, s_cd, s_da)* are lofted between the edges to skin the four faces of
the tower.

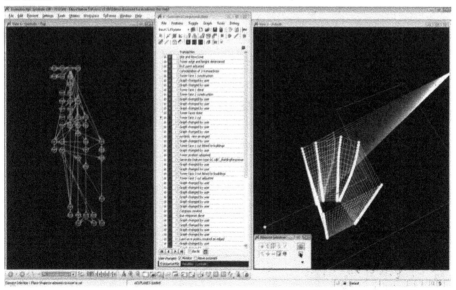

4.17 *Tower Formation* system making 10

```
transaction modelBased "Tower face 2 cut"
{
    feature Setback GC.GraphVariable
    {
        Value                    = 4.0;
        RangeMaximum             = 10.0;
    }
    feature c2a GC.BsplineCurve
    {
        Surface                  = s_bc;
        Parameter                = .4;
        Direction                = DirectionOption.V;
        SymbolXY                 = {103, 114};
    }
    feature c2b GC.BsplineCurve
    {
        FitPoints                = {p2b};
    }
    feature c2c GC.BsplineCurve
    {
        Surface                  = s_bc;
        Parameter                = .8;
        Direction                = DirectionOption.V;
        SymbolXY                 = {104, 114};
    }

    feature p2b GC.Point
    {
        CoordinateSystem         = baseCS;
        Xtranslation             = point16.X;
        Ytranslation             = point16.Y;
        Ztranslation             = p3.Z;
    }
    feature point16 GC.Point
    {
        Curve                    = 13;
        T                        = <free> (0.446836411407735);
        HandleDisplay            = DisplayOption.Display;
    }
    feature s_bc2 GC.BsplineSurface
    {
        Curves                                                  =
{c3[0],c2c,c2b[0],c2a,c222222[0]};
        UcurveDisplay            = 20;
        VcurveDisplay            = 20;
    }
}

transaction modelBased "Graph changed by user"
{
    feature point16 GC.Point
    {
        T                        = .3;
    }
}

    feature s_da GC.BsplineSurface
    {
        Curves                   = {c1[0],c4[0]};
        UcurveDisplay            = 20;
        VcurveDisplay            = 20;
    }
}
```

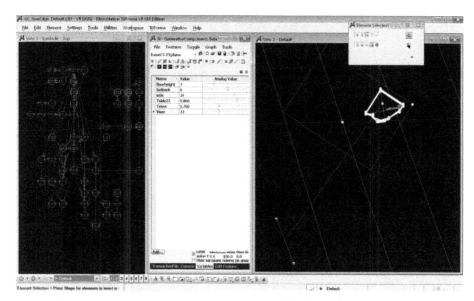

Response to buildings:

Rule: Tower should respond to the gaps between surrounding buildings that offer possibilities for open views. A V shape is cut into the tower face to maximise the façade area that faces the open view.

Points *building1* and *building2* are created referencing the corners of two existing buildings to the north of the tower.

A bisector connects the mid-distance between the points with the center of the tower.

```
transaction modelBased "Tower face 2 cut linked to buildings"
{
    feature Setback GC.GraphVariable
    {
        RangeMinimum              = 0.3;
        RangeMaximum              = 0.6;
        RangeStepSize             = 0.3;
    }
    feature Tside23 GC.GraphVariable
    {
        Value                     = 0.264;
        LimitValueToRange         = true;
        RangeMinimum              = 0.2;
        RangeMaximum              = 0.4;
        RangeStepSize             = 0.0;
    }
    feature building1 GC.Point
    {
        CoordinateSystem          = baseCS;
        Xtranslation              = <free> (235.039156550752);
        Ytranslation              = <free> (78.1398182662286);
        Ztranslation              = p3[0].Z;
        SymbolXY                  = {104, 102};
        HandleDisplay             = DisplayOption.Display;
    }
    feature building2 GC.Point
    {
        CoordinateSystem          = baseCS;
        Xtranslation              = <free> (240.123714996232);
        Ytranslation              = <free> (44.7562596406535);
        Ztranslation              = p3[0].Z;
        SymbolXY                  = {105, 102};
        HandleDisplay             = DisplayOption.Display;
    }
```

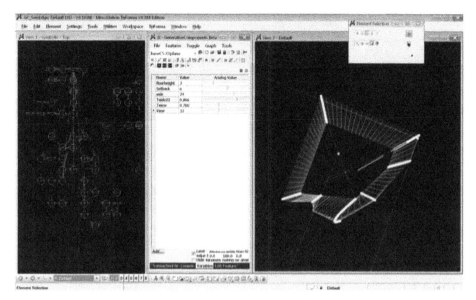

4.19 *Tower Formation* system making 12

A V shape cut is introduced at the intersection of the bisector and the tower's north face.

The isocurve that defines the V cut leans to the inside of the tower as it rises through the floors, giving greater view exposure at higher levels.

```
feature buildingtarget GC.Point
{
    Curve                  = line07;
    T                      = <free> (0.586405318050557);
    HandleDisplay          = DisplayOption.Display;
}
feature line05 GC.Line
{
    StartPoint             = building1;
    EndPoint               = building2;
    SymbolXY               = {106, 101};
    Display                = DisplayOption.Display;
}
feature line06 GC.Line
{
    StartPoint             = p3[0];
    EndPoint               = p1[0];
    SymbolXY               = {103, 105};
    Display                = DisplayOption.Display;
}
feature line07 GC.Line
{
    StartPoint             = p2[0];
    EndPoint               = p4[0];
    SymbolXY               = {103, 106};
    Display                = DisplayOption.Display;
}
feature line08 GC.Line
{
    StartPoint             = building1;
    EndPoint               = point10;
    SymbolXY               = {104, 103};
    Display                = DisplayOption.Hide;
}
feature line09 GC.Line
{
    StartPoint             = building2;
    EndPoint               = point10;
    SymbolXY               = {105, 103};
    Display                = DisplayOption.Hide;
}
feature line10 GC.Line
{
    StartPoint             = point08;
    EndPoint               = buildingtarget;
    SymbolXY               = {106, 103};
}
feature p2b GC.Point
{
    Xtranslation           = point12.X;
    Ytranslation           = point12.Y;
    SymbolXY               = {101, 111};
}
feature point08 GC.Point
{
    Curve                  = line05;
    T                      = .5;
    SymbolXY               = {106, 102};
    Display                = DisplayOption.Hide;
    HandleDisplay          = DisplayOption.Display;
}
feature point10 GC.Point
{
    Intersector0           = line06;
    Intersector1           = line07;
    SymbolXY               = {105, 105};
}
feature point12 GC.Point
{
    Intersector0           = line10;
    Intersector1           = 13;
}
feature point16 GC.Point
{
    T                      = Tside23;
}
}
```

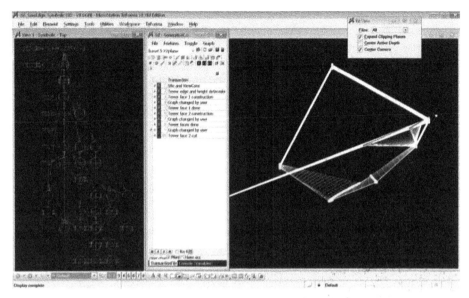

4.20 *Tower Formation* system making 13

Top view before the V cut in response to the in-between space.

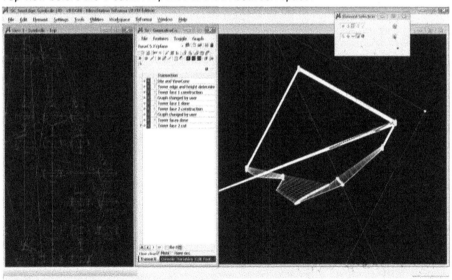

4.21 *Tower Formation* system making 14

Top view after the V cut in response to the in-between space.

```
transaction generateFeatureType "Generate feature type GC.UBC_BuildingResponse"
{
    type                    = GC.UBC_BuildingResponse;
    loadIntoFutureSessions  = true;
    inputProperties         = {
                        property GC.IPoint baseCS
                        {
                            feature                 = baseCS;
                            isReplicatable          = true;
                            isParentModel           = true;
                        }
                        property GC.IPoint building1
                        {
                            feature                 = building1;
                            isReplicatable          = true;
                        }
                        property GC.IPoint building2
                        {
                            feature                 = building2;
                            isReplicatable          = true;
                        }
                        property GC.BsplineCurve c222222
                        {
                            feature                 = c222222;
                            isOptional              = true;
                            isReplicatable          = true;
                        }
                        property GC.BsplineCurve c2a
                        {
                            feature                 = c2a;
                            isOptional              = true;
                            isReplicatable          = true;
                        }
                        property GC.BsplineCurve c2c
                        {
                            feature                 = c2c;
                            isOptional              = true;
                            isReplicatable          = true;
                        }
                        property GC.IPoint point10
                        {
                            feature                 = point10;
                            isOptional              = true;
                            isReplicatable          = true;
                        }
                };

                        property GC.Point p3
                        {
                            feature                 = p3;
                            isDynamic               = true;
                        }
                        property GC.Point point12
                        {
                            feature                 = point12;
                            isDynamic               = true;
                        }
                        property GC.BsplineSurface s_bc2
                        {
                            feature                 = s_bc2;
                            isDynamic               = true;
                        }
                };
}
```

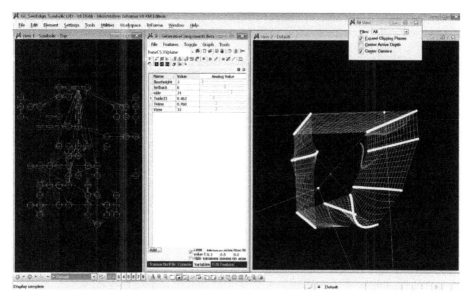

4.22 *Tower Formation* system making 15

Building Response Feature generation:

UBC_BuildingResponse is a user-defined *GenerativeComponents* feature for the building response to in-between views generated as part of the software interface. The response feature can be applied to any of the faces of the tower depending on the input values of the coordinates of the surrounding buildings.

The feature *UBC_BuildingResponse* is used to cut a similar V shape into the east face of the tower. The east face responds to two buildings adjacent to the site. The V cut width is directly proportional to the open view width between the two buildings.

103

```
outputProperties        = {
                        property GC.BsplineCurve c2b
                        {
                            feature              = c2b;
                            isDynamic            = true;
                        }
                        property GC.BsplineCurve c3
                        {

                            feature              = c3;
                            isDynamic            = true;
                        }
                        property GC.Line l3
                        {
                            feature              = l3;
                            isDynamic            = true;
                        }

                        property GC.Line line05
                        {
                            feature              = line05;
                            isDynamic            = true;
                        }
                        property GC.Line line10
                        {
                            feature              = line10;
                            isDynamic            = true;
                        }
                        property GC.Point p2b
                        {
                            feature              = p2b;
                            isDynamic            = true;
                        }
    internalProperties       = {
                        property GC.Line line08
                        {
                            feature              = line08;
                            isDynamic            = true;
                        }
                        property GC.Line line09
                        {
                            feature              = line09;
                            isDynamic            = true;
                        }
                        property GC.Point point08
                        {
                            feature              = point08;
                            isDynamic            = true;
                        };
}
```

```
transaction modelBased "North"
{
    feature Tnorth GC.GraphVariable
    {
        Value                   = 0.25;
        LimitValueToRange       = true;
        RangeMaximum            = 1.0;
        RangeStepSize           = 0.0;
        SymbolXY                = {87, 101};
    }
    feature bsplineCurve03 GC.BsplineCurve
    {
        Parameter               = Tside34+.5;
    }
    feature circle01 GC.Circle
    {
        CenterPoint             = point20;
        Radius                  = 5;
        Support                 = baseCS.XYplane;
        SymbolXY                = {90, 100};
    }
    feature east GC.Point
    {
        Curve                   = circle01;
        T                       = north.T-.25;
        SymbolXY                = {90, 103};
        HandleDisplay           = DisplayOption.Display;
    }
    feature line15 GC.Line
    {
        StartPoint              = point20;
        Direction               = northdirection;
        Length                  = circle01.Radius+2;
        SymbolXY                = {91, 101};
    }
    feature north GC.Point
    {
        Curve                   = circle01;
        T                       = Tnorth;
        SymbolXY                = {90, 101};
        HandleDisplay           = DisplayOption.Display;
    }
    feature northdirection GC.Direction
    {
        OriginPoint             = point20;
        DirectionPoint          = north;
        SymbolXY                = {90, 102};
    }
    feature text01 GC.Text
    {
        Placement               = line15.EndPoint;
        TextString              = " North ";
        SymbolXY                = {91, 103};
    }
}

transaction modelBased "Compass created"
{
    feature direction01 GC.Direction
    {
        OriginPoint             = point20;
        DirectionPoint          = east;
        SymbolXY                = {90, 104};
    }
}
transaction modelBased "Sun response done"
{
    feature Tnorth GC.GraphVariable
    {
        Value                   = 0.165;
        RangeMaximum            = 0.25;
        SymbolXY                = {89, 101};
    }
}
```

105

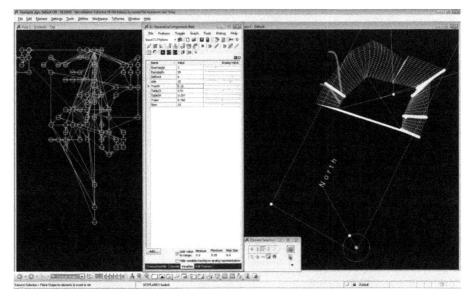

4.23 *Tower Formation* system making 16

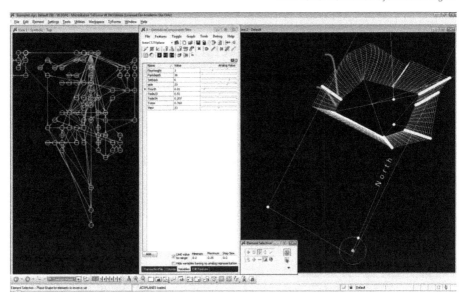

4.24 *Tower Formation* system making 17

```
feature Tside23 GC.GraphVariable
    {
        Value                    = 0.51;
    }
    feature line15 GC.Line
    {
        Length                   = circle01.Radius+70;
    }
    feature p3 GC.Point
    {
        Ytranslation             = p1.Y-side-Abs(north.X);
    }
    feature p4 GC.Point
    {
        Xtranslation             = p1.X+Abs(north.Y);
        Ytranslation             = p1.Y-side-Abs(north.Y);
    }
    feature point03 GC.Point
    {
        Display                  = DisplayOption.Hide;
    }
    feature point20 GC.Point
    {
        Xtranslation             = point03.X;
        Ytranslation             = point03.Y;
    }
}
```

Response to Southern exposure:

Rule: The tower should respond to the north direction in order to maximise southern exposure.

Tnorth is a variable that controls the north direction relative to the site. It works by defining the position of point *north* relative to its displacement over the circle (compass).

The X and Y translation of edges *p3* and *p4* are linked to the north direction. The translation changes relative to the original position to allow more south exposure on the building's faces. The first screenshot shows the real north direction relative to the site, while both screenshots together show the response of the edges to a change in the north direction (or a change in the tower's orientation). (Figures 4.23, 4.24)

```
transaction modelBased "Response to park"
{
    feature line17 GC.Line
    {
        StartPoint                    = point33;
        EndPoint                      = p1;
    }
    feature p4 GC.Point
    {
        Xtranslation                                                  =
p1.X+Abs(north.Y)+line17[0].Length*.1-line17.Length*.1;
    }
    feature point32 GC.Point
    {
        CoordinateSystem              = baseCS;
        Xtranslation                  = point04.X+Setback;
        Ytranslation                  = point04.Y-Setback;
        Ztranslation                  = point04.Z;
        HandleDisplay                 = DisplayOption.Display;
    }
    feature point33 GC.Point
    {
        CoordinateSystem              = baseCS;
        Xtranslation                  = point04.X+Setback;
        Ytranslation                  = point04.Y-Setback-side;
        Ztranslation                  = point04.Z;
    }
    feature point34 GC.Point
    {
        CoordinateSystem              = baseCS;
        Xtranslation                  = point04.X+3*Setback;
        Ytranslation                  = point04.Y-Setback-side;
        Ztranslation                  = point04.Z+2;
    }
    feature point35 GC.Point
    {
        CoordinateSystem              = baseCS;
        Xtranslation                  = point04.X+6*Setback;
        Ytranslation                  = point04.Y-Setback-side;
        Ztranslation                  = point04.Z+1;
    }
}
```

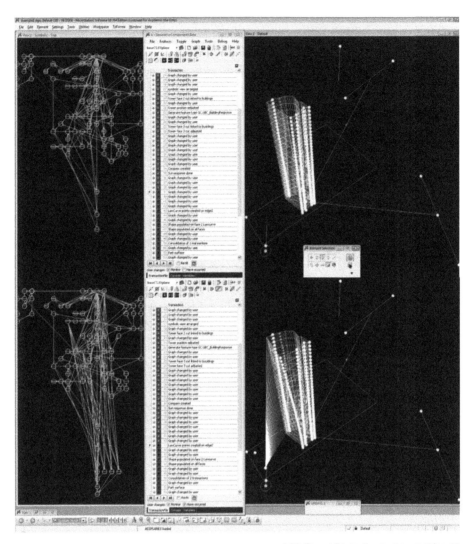

Response to Park:

Rule: The tower configuration promotes maximum exposure to the park of the building and its units.

The x translation of the corner point series *p4* is adjusted to respond to the park. A point *point33* created at the edge of the park is linked with line *line17* to the face of the tower. The X translation of *p4* is directly related to the length of the connecting line *line17*.

110

```
transaction modelBased "LawCurve points created on edge2"
{
    feature LawC1 GC.Point
    {
        CoordinateSystem          = coordinateSystem02;
        Xtranslation              = 0;
        Ytranslation              = 0;
        Ztranslation              = <free> (0.0);
        HandleDisplay             = DisplayOption.Display;
    }
    feature LawC2 GC.Point
    {
        CoordinateSystem          = coordinateSystem02;
        Xtranslation              = <free> (9.6936374195728);
        Ytranslation              = <free> (37.6503328615144);
        Ztranslation              = <free> (0.0);
        HandleDisplay             = DisplayOption.Display;
    }
    feature LawC3 GC.Point
    {
        CoordinateSystem          = coordinateSystem02;
        Xtranslation              = lawCurveFrame01.Xdimension;
        Ytranslation              = lawCurveFrame01.Ydimension;
        Ztranslation              = <free> (0.0);
        HandleDisplay             = DisplayOption.Display;
    }
    feature bsplineCurve05 GC.BsplineCurve
    {
        FitPoints                 = {point05,point07,building1};
        SymbolXY                  = {89, 107};
    }
    feature coordinateSystem02 GC.CoordinateSystem
    {
        Model                     = "LawCurve";
    }
    feature lawCurve01 GC.LawCurve
    {
        LawCurveFrame             = lawCurveFrame01;
        ControlPoints             = {LawC1,LawC2,LawC3};
        CurveControl              = CurveOption.ByPoints;
        Order                     = 4;
        IndependentVariable       = Series(0,20,1);
    }
    feature lawCurveFrame01 GC.LawCurveFrame
    {
        Plane                     = coordinateSystem02.XYplane;
        Xdimension                = 20;
        Ydimension                = c222222[0].Length;
        Xaxis                     = Series(0,20,1);
        Yaxis                     = Series(0,c222222[0].Length,1);
    }
    feature p2_new GC.Point
    {
        Curve                     = c222222[0];
        DistanceAlongCurve        = lawCurve01.DependentVariable;
    }
}
```

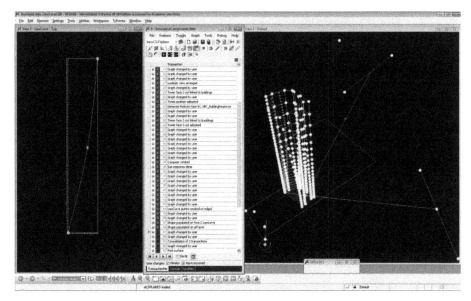

Law Curve:

Rule: The glazing-to-solid ratio should increase at higher levels of the building to make use of view exposures. At the same instance, openings at the base levels should be bigger to compensate for the decrease in lighting conditions.

A law curve controls the spacing of points along the tower edges. The law curve is set on a graph in which:

1. The X-axis (independent) represents the number of points along the curve.
2. The Y-axis (dependent) represents the spacing between the points.

The law curve has control points *(LawC1, Lawc2, Lawc3)* that change its curvature, and therefore change the gradient spacing of points along the curve (Y).

Example: if the curve is a straight line, the spacing between the points (Y) is equal.

```
transaction modelBased "LawCurve points created on edge2"
{
    feature LawC1 GC.Point
    {
        CoordinateSystem          = coordinateSystem02;
        Xtranslation              = 0;
        Ytranslation              = 0;
        Ztranslation              = <free> (0.0);
        HandleDisplay             = DisplayOption.Display;
    }
    feature LawC2 GC.Point
    {
        CoordinateSystem          = coordinateSystem02;
        Xtranslation              = <free> (9.6936374195728);
        Ytranslation              = <free> (37.6503328615144);
        Ztranslation              = <free> (0.0);
        HandleDisplay             = DisplayOption.Display;
    }
    feature LawC3 GC.Point
    {
        CoordinateSystem          = coordinateSystem02;
        Xtranslation              = lawCurveFrame01.Xdimension;
        Ytranslation              = lawCurveFrame01.Ydimension;
        Ztranslation              = <free> (0.0);
        HandleDisplay             = DisplayOption.Display;
    }
    feature bsplineCurve05 GC.BsplineCurve
    {
        FitPoints                 = {point05,point07,building1};
        SymbolXY                  = {89, 107};
    }
    feature coordinateSystem02 GC.CoordinateSystem
    {
        Model                     = "LawCurve";
    }
    feature lawCurve01 GC.LawCurve
    {
        LawCurveFrame             = lawCurveFrame01;
        ControlPoints             = {LawC1,LawC2,LawC3};
        CurveControl              = CurveOption.ByPoints;
        Order                     = 4;
        IndependentVariable       = Series(0,20,1);
    }
    feature lawCurveFrame01 GC.LawCurveFrame
    {
        Plane                     = coordinateSystem02.XYplane;
        Xdimension                = 20;
        Ydimension                = c222222[0].Length;
        Xaxis                     = Series(0,20,1);
        Yaxis                     = Series(0,c222222[0].Length,1);
    }
    feature p2_new GC.Point
    {
        Curve                     = c222222[0];
        DistanceAlongCurve        = lawCurve01.DependentVariable;
    }
}
```

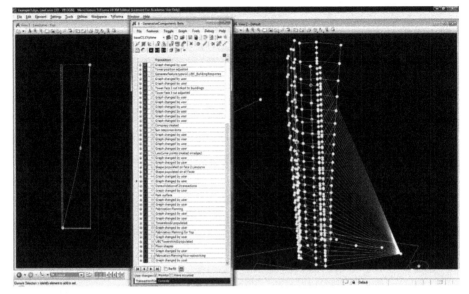

4.27 *Tower Formation* system making 20

Points on the tower's edges and isocurves with reference to the law curve's Y axis (spacing between instances of each point) are created.

A rectangular polygon *shape01* is created on top of these points by joining them. The polygons can be used as a base to populate skin design components to the tower's surface, while responding to the varying dimensions of each individual polygon.

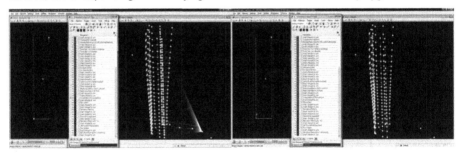

4.28 *Tower Formation* system making 21

Adjustments to the control point of the law curve correspond to a change in Y translation of the polygon points. Four control points manipulate the law curve in order to change the spacing of points at the top and bottom levels of the tower.

114

```
transaction modelBased "Fabrication Planning"
{

    feature coordinateSystem03 GC.CoordinateSystem
    {
        CoordinateSystem        = baseCS;
        Xtranslation            = 0;
        Ytranslation            = 200;
        Ztranslation            = 0;
    }
    feature export01 GC.Export
    {
        FeaturesToExport        = fabricationPlanning01;
        ExportDesignFile        = 1;
        ExportSeedFile          = 2;
    }
    feature fabricationPlanning01 GC.FabricationPlanning
    {
        CoordinateSystem        = coordinateSystem03;
        Shapes                  = shape01;
        Xspacing                = 20;
        Yspacing                = 1;
        Fill                    = true;
        TextStyle               = textStyle01;
    }
}
```

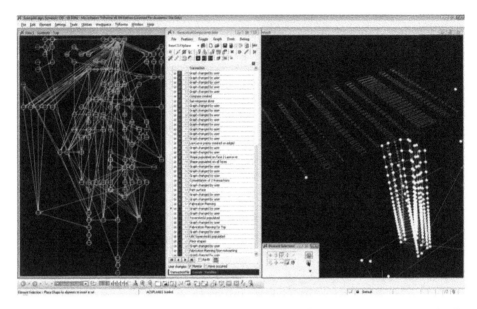

Fabrication Planning:

Fabrication Planning is a feature in *GenerativeComponents* that can prepare geometry for digital fabrication. *Shape01* (the tower's skin) with its varying instances is laid out on a XY plane. This polygon layout is dynamically linked to changes to the tower's geometry.

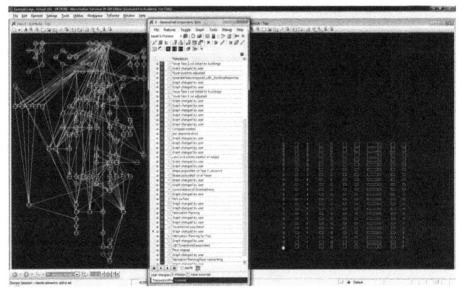

```
transaction modelBased "Towerskin02 populated"
{
    feature Rcomponent GC.GraphVariable
    {
        Value                    = .2;
        LimitValueToRange        = true;
        RangeMinimum             = 0.1;
        RangeMaximum             = 0.6;
        RangeStepSize            = 0.1;
    }
    feature Tnumber GC.GraphVariable
    {
        Value                    = .333333;
    }
    feature p4 GC.Point
    {
        Display                  = DisplayOption.Hide;
    }
    feature shape01 GC.Shape
    {
        Display                  = DisplayOption.Hide;
    }
    feature uBCTowerskin0201 GC.UBCTowerskin02
    {
        Rcomponent               = Rcomponent;
        Tnumber                  = Tnumber;
        shape01                  = shape01;
    }
    feature uBCViewCone01 GC.UBCViewCone
    {
        Display                  = DisplayOption.Hide;
    }
}




transaction modelBased "UBCTowerskin03 populated"
{
    feature uBCTowerskin0301 GC.UBCTowerskin03
    {
        Rcomponent               = Rcomponent;
        Tnumber                  = Tnumber;
        shape01                  = shape01;
        SymbolXY                 = {103, 126};
        Display                  = DisplayOption.Display;
    }
}
```

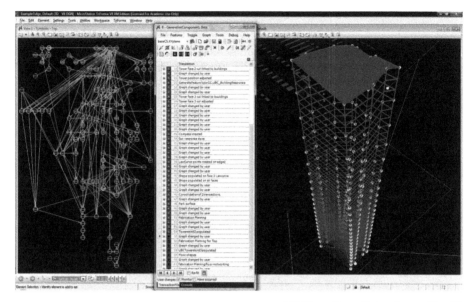

The polygon *Shape01* is used to populate facade components across the building. The adjustable spacing of the polygon points allows matching the component to the instances of the polygon. In this study, the component divides each polygon into 3 horizontal panels, one solid and two glazed panels. Glazing sizes vary depending on levels of the tower consistent with the glazing-to-solid ratio rule.

```
transaction modelBased "Floor shapes"
{
    deleteFeature fabricationPlanning03;
    feature shape05 GC.Shape
    {
        Vertices                                                    =
{point27[18],point28[18],point29[18],point23[18],point21[18],point22[18],point19[18
],p2_new[18],point24[18],point25[18],point26[18]};
        SymbolXY                  = {105, 121};
    }
    feature shape06 GC.Shape
    {
        Vertices                                                    =
{point27[12],point28[12],point29[12],point23[12],point21[12],point22[12],point19[12
],p2_new[12],point24[12],point25[12],point26[12]};
        SymbolXY                  = {106, 121};
    }
    feature shape07 GC.Shape
    {
        Vertices                                                    =
{point27[6],point28[6],point29[6],point23[6],point21[6],point22[6],point19[6],p2_ne
w[6],point24[6],point25[6],point26[6]};
        SymbolXY                  = {107, 121};
    }
    feature textStyle01 GC.TextStyle
    {
        SymbolXY                  = {101, 126};
    }
}

transaction modelBased "Graph changed by user"
{
    feature shape06 GC.Shape
    {
        Vertices                                                    =
{point27[14],point28[14],point29[14],point23[14],point21[14],point22[14],point19[14
],p2_new[14],point24[14],point25[14],point26[14]};
    }
}
```

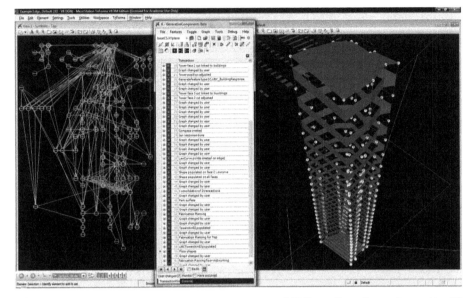

4.32 *Tower Formation* system making 25

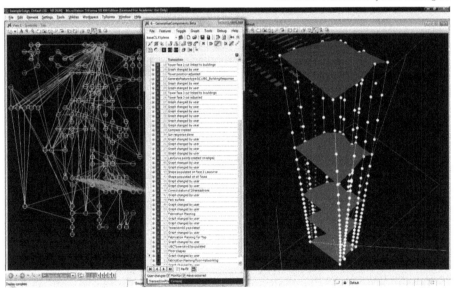

4.33 *Tower Formation* system making 26

Floor slabs *shape04*, *shape05*, *shape06*, and *shape07* are created in preparation of the laser cutting process.

4.34 Physical model
for *Tower Formation*

Comments

The design model of a high-rise residential building illustrates the integration of procedural thinking and computation into the design process. Several interrelated parameters define how the building responds to its context. References for the design process are abstracted and incorporated into a set of rules that are executed by the software. As a result, the developed design system is dynamically responsive to the contextual parameters. The design model mediates and represents the complex interaction of all parameters.

A similar design system can be applied to building designs at other sites. Equally, additional parameters can be incorporated to add features such as shading and reflecting devices in response to solar movement.

As with the previous exploration of domestic living spaces, form in *Tower Formation* is a result of the designed responsive system. The set rules include the design limitations needed to control the performance of the system. Form is a product of the design process.

Tower Formation illustrates spatial configuration and form as a result of a rational design methodology. The building is designed through the strict execution of procedures specifically set to control the building's response to contextual references. Subjective design considerations can be included into the design process through the rules that govern the formation process. The selection of parameters reflects intentions of the designer.

---

[v] Vancouver Views: Council Approved View Cones. 18 10 2008
<http://vancouver.ca/commsvcs/views/viewcones/91.htm>.

5. Shade

Introduction

Shade explores concepts for a responsive sun shading system that uses parametric and computational techniques to change its configuration by strictly following an explicit programmatic function. *Shade* also illustrates how rule-based procedures can be utilised to add functional patterns to the façade of a building.

Design Approach

A rule-based design approach- similar to previous explorations- is used to generate programmatic and formal complexity out of a simple state. First, design elements are abstracted into simple geometries. The geometries are then integrated into the parametric system. Second, the design program is translated into a set of rules that guide the performance of geometries within the parametric system. Rules dictate how the system responds to changing parameters. Third, changes of outside influences trigger a dynamic response in the system, guaranteed by the strict execution of the set rules. At any instance of the model, form emerges as a result of the responsiveness of the design system to changing parameters.

The parametric design system addresses the orientation of the building façade. Depending on the cordial directions (orientation), the design system prefaces horizontal or vertical overhangs of a shading system.

Design Background

Inspired by Le Corbusier's modernist design of egg-crate like shading overhangs that he used in a number of his projects, *Shade* investigates how parametric modeling can be used to introduce responsive shading overhangs. For the Unite d' Habitation, Le Corbusier used a modular concrete screen wall to reduce solar heat gain. The screen wall is composed of horizontal and vertical shading elements

5.1 Unite d' Habitation, Marseilles, France. 1952

with consistent projection lengths. While the screen walls that combines the characteristics of both horizontal and vertical shading elements offers a considerable potential for solar control. The use of shading elements of similar depth questions the efficiency of such a system to provide the appropriate shading on each of the glazing surfaces.

Position

Vertical fins of a sun shading system are generally assumed appropriate for providing solar control on East and West facades, while horizontal overhangs provide the best solar control on south facing facades. Adjustability of both horizontal and vertical components maximises the positive effect of shading devices.

Objective

The parametric design for the shading system is designed to take into consideration facade orientation and to govern the appropriate lengths of horizontal and vertical shading elements to be added to the wall as a screen system. The design system responds dynamically to changes in its position relative to the north direction.

Design process

The design process starts with abstracting an exterior wall into a rectangle.

A series of overhangs are then added to the base rectangle. The maximum depth of the overhang $Lmax$ is dependent on the number of subdivisions and therefore the subdivision length S. Where:

$$Lmax = S$$

The actual length of the overhang L depends on its position in relation to the north direction. A factor f, ranging from 0 to 1, correlates to the position relative to the north direction. Factor f is multiplied by the maximum length to generate the actual length. Where:

$$L = Lmax * f$$

Influencing Factors:

1. North direction
2. Thickness of overhangs (constant)

Rules:

1. North facing walls: No shading Overhangs $(f=0)$.
2. South facing walls: The horizontal overhangs will have the maximum allowed length $(f=1)$; and there will be no vertical overhangs $(f= 0)$.
3. East + West facing walls: The vertical overhangs will have the maximum allowed length $(f=1)$; and there will be no horizontal overhangs $(f= 0)$.

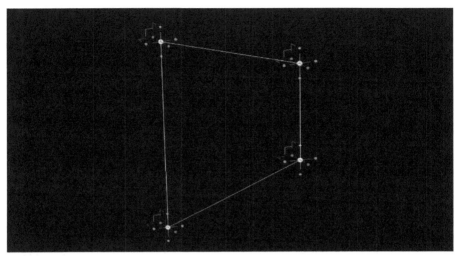

System making

The following pages illustrate the process of building the system and integrating the design constraints in *GenerativeComponents*.

126

```
transaction modelBased "points and shape"
{
    feature point01 GC.Point
    {
        CoordinateSystem          = baseCS;
        Xtranslation              = <free> (0.900256062153973);
        Ytranslation              = <free> (-1.17106949772331);
        Ztranslation              = <free> (-2.77555756156289E-17);
        Display                   = DisplayOption.Display;
        HandleDisplay             = DisplayOption.Display;
    }
    feature point02 GC.Point
    {
        CoordinateSystem          = baseCS;
        Xtranslation              = <free> (0.647841114966038);
        Ytranslation              = <free> (-1.6709146375336);
        Ztranslation              = <free> (6.06229618060974);
        Display                   = DisplayOption.Display;
        HandleDisplay             = DisplayOption.Display;
    }
    feature point03 GC.Point
    {
        CoordinateSystem          = baseCS;
        Xtranslation              = <free> (0.786013913746437);
        Ytranslation              = <free> (4.64199706144227);
        Ztranslation              = <free> (5.1023700547652);
        Display                   = DisplayOption.Display;
        HandleDisplay             = DisplayOption.Display;
    }
    feature point04 GC.Point
    {
        CoordinateSystem          = baseCS;
        Xtranslation              = <free> (0.576798427016679);
        Ytranslation              = <free> (4.52586903255925);
        Ztranslation              = <free> (0.212020244180056);
        Display                   = DisplayOption.Display;
        HandleDisplay             = DisplayOption.Display;
    }
    feature shape01 GC.Shape
    {
        Vertices                  = {point01,point02,point03,point04};
    }
}


transaction modelBased "2 vt lines"
{
    feature divisions GC.GraphVariable
    {
        Value                     = 4.0;
        LimitValueToRange         = true;
        RangeMinimum              = 1.0;
        RangeMaximum              = 4.0;
        RangeStepSize             = 1.0;
        SymbolXY                  = {99, 103};
    }
    feature line01 GC.Line
    {
        StartPoint                = shape01.Vertices[2];
        EndPoint                  = shape01.Vertices[3];
        Display                   = DisplayOption.Display;
    }
    feature line02 GC.Line
    {
        StartPoint                = shape01.Vertices[0];
        EndPoint                  = shape01.Vertices[1];
        Display                   = DisplayOption.Display;
    }
}
```

127

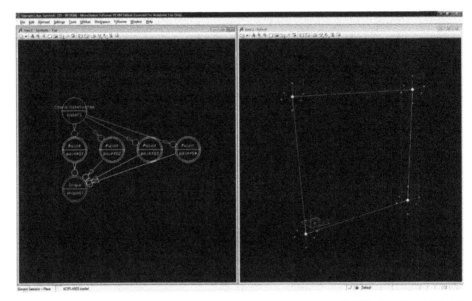

5.4 *Shade* system making 1

First, a rectangle is drawn on 4 points as a base for the system.

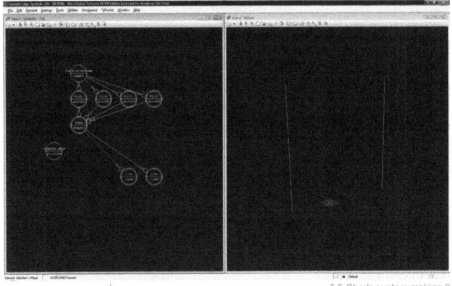

5.5 *Shade* system making 2

Vertical lines are drawn on the sides of the shape as a base for the subdivisions.

128

```
transaction modelBased "lines from point"
{
     feature point05 GC.Point
     {
         Curve                   = line02;
         T                       = Series(0,1,1/divisions);
         SymbolXY                = (104, 105);
         HandleDisplay           = DisplayOption.Display;
     }
     feature point06 GC.Point
     {
         Curve                   = line02;
         T                       = <free> (0.630790283878103);
         SymbolXY                = (103, 105);
         HandleDisplay           = DisplayOption.Display;
     }
     feature coordinateSystem01 GC.CoordinateSystem
     {
         Origin                  = shape01.Vertices[0];
         PrimaryDirection        = line03;
         PrimaryAxis             = AxisOption.Y;
         SecondaryDirection      = line02;
         SecondaryAxis           = AxisOption.Z;
     }
     feature line03 GC.Line
     {
         StartPoint              = shape01.Vertices[0];
         EndPoint                = shape01.Vertices[3];
         SymbolXY                = (100, 104);
     }
     feature line04 GC.Line
     {
         StartPoint              = point06;
         Direction               = coordinateSystem01.Xdirection;
         Length                  = .5;
         Color                   = 5;
     }
     feature line05 GC.Line
     {
         StartPoint              = point06;
         Direction               = baseCS.Xdirection;
         Length                  = .5;
         SymbolXY                = (103, 108);
         SymbolicModelDisplay    = null;
         Color                   = 5;
         FillColor               = -1;
         Free                    = true;
         Level                   = 0;
         LevelName               = "Default";
         LineStyle               = 0;
         LineStyleName           = "0";
         LineWeight              = 0;
         MaximumReplication      = true;
         RoleInExampleGraph      = null;
         Transparency            = 1;
     }
     feature CosAngle GC.GraphVariable
     {
         Value
Cos(Angle(point06,line05.EndPoint,line04.EndPoint));
         SymbolXY                = (103, 109);
     }
     feature LineLength GC.Line
     {
         StartPoint              = point06;
         Direction               = coordinateSystem01.Xdirection;
         Length                  = CosAngle;
         SymbolXY                = (102, 108);
         SymbolicModelDisplay    = null;
         Color                   = 0;
         FillColor               = -1;
         Free                    = true;
         Level                   = 0;
         LevelName               = "Default";
         LineStyle               = 0;
         LineStyleName           = "0";
         LineWeight              = 0;
         MaximumReplication      = true;
         RoleInExampleGraph      = null;
         Transparency            = 1;
     }
}
```

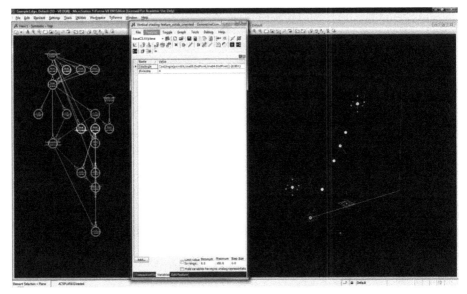

Point series *point05* is drawn to mark the division points on the vertical line. The spacing is controlled by the variable *divisions*.

A coordinate system is set on the vertical line to determine its orientation relative to the north *Xdirection*. Two directional lines *Line 04* and *Line05* start from a point on the vertical line, *Line04* is oriented in the default north direction and *Line05* is oriented perpendicular to the vertical line.

The variable CosAngle is set to determine the cosine of the angle between *Line04* and *Line 05*, which is then used as a factor to control the length of the overhang.

```
transaction modelBased "projection line"
{

feature proj GC.Point
    {
        Plane                       = baseCS.XYplane;
        PointToProjectOnToPlane     = point06;
        SymbolXY                    = {103, 106};
    }
    feature referencepoint GC.Point
    {
        CoordinateSystem            = baseCS;
        Xtranslation                = <free> (-5.42881925419483);
        Ytranslation                = <free> (-1.00912222452869);
        Ztranslation                = <free> (0.0);
        SymbolXY                    = {102, 105};
        HandleDisplay               = DisplayOption.Display;
    }
}

    feature line07 GC.Line
    {
        StartPoint                  = referencepoint;
        EndPoint                    = proj;
    }
}

transaction modelBased "shadeV2"
{
    feature LineLength GC.Line
    {
        Length                      = CosAngle+line07.Length*shadeV2;
    }
    feature shadeV2 GC.GraphVariable
    {
        Value                       = 0.08;
        LimitValueToRange           = true;
        RangeMaximum                = 1.0;
        RangeStepSize               = 0.0;
        SymbolXY                    = {99, 108};
    }
}

transaction modelBased "Graph changed by user"
{
    feature point08 GC.Point
    {
        Curve                       = line01;
        T                           = Series(0,1,1/divisions);
        SymbolXY                    = {99, 105};
        HandleDisplay               = DisplayOption.Display;
    }
}

transaction   modelBased   "Prepare   to   generate   feature   type
GC.UBC_LinebyCosAngleLength08"
{
    feature CosAngle GC.GraphVariable
    {
        Value                                                        =
Cos(Angle(point06,line05.EndPoint,point07));
    }
    feature plane01 GC.Plane
    {
        CoordinateSystem            = baseCS;
        Xtranslation                = point06.X;
        Ytranslation                = point06.Y;
        Ztranslation                = point06.Z;
    }
    feature point07 GC.Point
    {
        Plane                       = plane01;
        PointToProjectOnToPlane     = line04.EndPoint;
    }
}
```

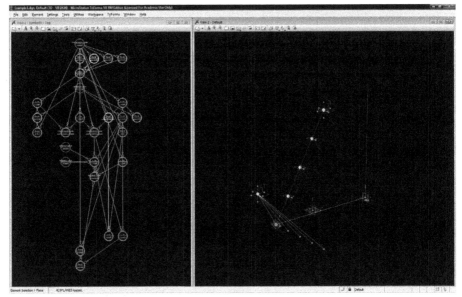

5.7 *Shade* system making 4

To determine the planer angle between the directional lines *Line 04* and *Line05*, a projection of *Line 04* is constructed to the horizontal plane. Variable *CosAngle* is adjusted to measure the cosine of the angle between two directional lines that lie on the same horizontal plane.

The division point series *point06* is projected on the default horizontal plane, and linked to reference point on the same plane. The distance between the projection points and the reference point *line07.Length* is used to vary the length of the overhangs. When the sideline is vertical, the overhangs have the same lengths. When the sideline is on an incline, the resulting overhangs have varying lengths because of their varying orientations.

132

```
transaction    modelBased    "Prepare    to    generate    feature    type
GC.UBC_LinebyCosProjectedAngleLength09"
{
    deleteFeature LineLength;
    feature Lmax GC.GraphVariable
    {
        Value                   = .5;
        LimitValueToRange       = true;
        RangeMaximum            = 3.0;
        RangeStepSize           = 0.0;
        SymbolXY                = {99, 107};
    }
    feature L GC.Line
    {
        StartPoint              = point06;
        Direction               = coordinateSystem01.Xdirection;
        Length                  = CosAngle*Lmax+line07.Length*shadeV2;
        SymbolXY                = {101, 108};
    }

    feature coordinateSystem02 GC.CoordinateSystem
    {
        Origin                  = shape01.Vertices[3];
        PrimaryDirection        = line03;
        PrimaryAxis             = AxisOption.Y;
        SecondaryDirection      = line01;
        SecondaryAxis           = AxisOption.Z;
        SymbolXY                = {99, 106};
    }
    feature point06 GC.Point
    {
        T                       = Series(0,1,1/divisions);
    }
    feature proj2 GC.Point
    {
        Plane                   = baseCS.XYplane;
        PointToProjectOnToPlane = point08;
        SymbolXY                = {97, 106};
    }
}
```

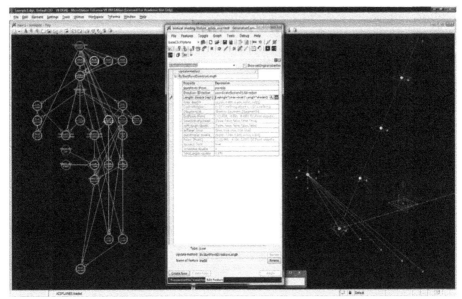

Calculation of the length of an overhang:

$L = Lmax * CosAngle$ (CosineAngle) $+ line07.Length* 0.1$ (length of projection line)

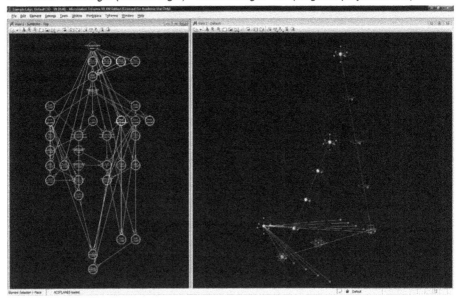

Divisions and overhang directional lengths are constructed on the second sideline using the same sequence.

134

```
transaction modelBased "surfaces between lines -south wall works"
{
    feature L GC.Line
    {
        Length                                                          =
Abs(CosAngle*Lmax+line07.Length*shadeV2);
    }

    feature bsplineSurface05 GC.BsplineSurface
    {
        StartCurve                 = line10;
        EndCurve                   = L;
        Color                      = 3;
    }
    feature divisions GC.GraphVariable
    {
        Value                      = 7.0;
        RangeMaximum               = 10.0;
    }
}
```

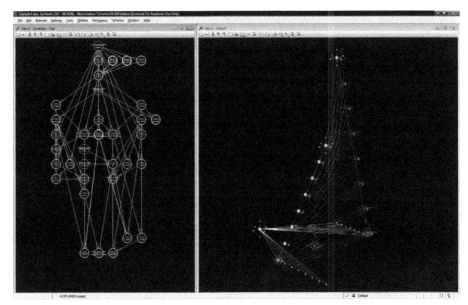

Surfaces are constructed between the overhang length lines on the sidelines.

```
transaction modelBased "East facing wall test"
{

    feature baseCS GC.CoordinateSystem
    {
        Display                  = DisplayOption.Display;
    }
    feature line10 GC.Line
    {
        Length                                              =
Abs(cosAngle2*Lmax+line11.Length*shadeV2);
    }
    feature point03 GC.Point
    {
        Ytranslation             = <free> (-4.34766465394409);
    }
    feature point04 GC.Point
    {
        Ytranslation             = <free> (-4.17338362707951);
    }
    feature shadeV2 GC.GraphVariable
    {
        Value                    = 0.0;
    }
}

transaction modelBased "South facing wall test"
{
    feature point02 GC.Point
    {
        Xtranslation             = <free> (1.36936699571921);
        Ytranslation             = <free> (-4.67742589723736);
    }
    feature point03 GC.Point
    {
        Xtranslation             = <free> (2.55293201974526);
        Ytranslation             = <free> (-11.8859752799482);
    }
    feature point04 GC.Point
    {
        Xtranslation             = <free> (2.48515215470223);
        Ytranslation             = <free> (-10.1328428456147);
    }
}
transaction modelBased "length equation changed"
{
    feature L GC.Line
    {
        Length                                              =
Abs(CosAngle)*Lmax+line07.Length*shadeV2*Abs(CosAngle);
    }

    feature louverthickness GC.GraphVariable
    {
        Value                    = .15;
    }
    feature solid01 GC.Solid
    {
        SurfaceToOffset          = bsplineSurface05;
        OffsetAboveSurface       = 0;
        OffsetBelowSurface       = louverthickness;
        Color                    = 3;
    }
}

transaction modelBased "solids"
{
    feature solid01 GC.Solid
    {
        OffsetAboveSurface       = louverthickness;
        OffsetBelowSurface       = 0;
    }

}
```

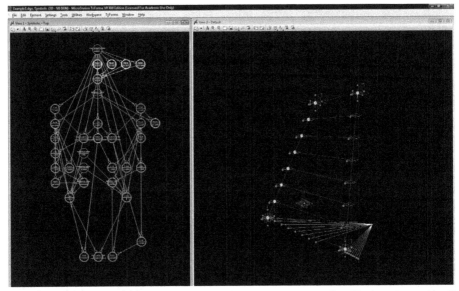

5.11 *Shade* system making 8

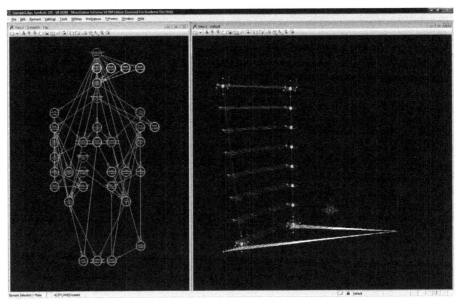

5.12 *Shade* system making 9

Changes to the wall orientation illustrate how the horizontal overhang system responds to the east direction with minimal overhangs (Figure 5.11), and to south orientation with maximum overhang lengths. (Figure 5.12)

138

```
transaction generateFeatureType "Generate feature type GC.UBC_VerticalShading07"
{
    type                    = GC.UBC_VerticalShading07;
    loadIntoFutureSessions  = true;
    inputProperties         = {
                            property GC.IPoint baseCS
                            {
                                feature             = baseCS;
                                isReplicatable      = true;
                                isParentModel       = true;
                            }
                            property GC.Shape shape01
                            {
                                feature             = shape01;
                            }
                            property GC.IPoint referencepoint
                            {
                                feature             = referencepoint;
                            }
                            property double louverthickness
                            {
                                feature             = louverthickness;
                                isReplicatable      = true;
                            }
                            property double divisions
                            {
                                feature             = divisions;
                                isReplicatable      = true;
                            }
                            property double Lmax
                            {
                                feature             = Lmax;
                            }
                            };
    outputProperties        = {

                            property GC.Solid solid01
                            {
                                feature             = solid01;
                                isDynamic           = true;
                            }
                            };
    internalProperties      = {
                            property GC.BsplineSurface bsplineSurface05
                            {
                                feature             =
bsplineSurface05;
                                isConstruction      = true;
                                isDynamic           = true;
                            }

                            property GC.GraphVariable CosAngle
                            {
                                feature             = CosAngle;
                                isConstruction      = true;
                                isDynamic           = true;
                            }
                            property GC.GraphVariable cosAngle2
                            {
                                feature             = CosAngle2;
                                isConstruction      = true;
                                isDynamic           = true;
                            }
                            };
}
```

139

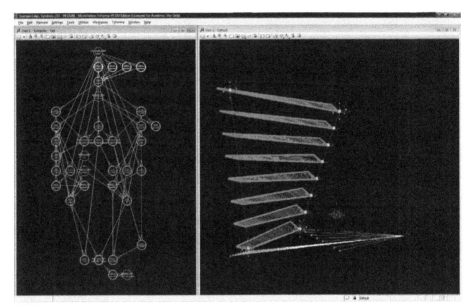

Solids are constructed on the overhang surfaces. The thickness of the overhang is manipulated by the variable *thickness*.

A user-defined shading feature *GC.UBC_VerticalShading07* is generated and stored inside the default features list of *GenerativeComponents*. This feature stores the process of designing the horizontal shading system including the relationships of parameters.

The generation of feature *GC.UBC_VerticalShading07* enables the model of the shading system to be applied to any configuration in the future, providing the base shape and the input variables are specified.

Inputs:

1. Base coordinate system
2. Base shape
3. Reference point
4. Louver thickness
5. Number of divisions
6. Maximum Louver Length

Outputs:

1. Louver configuration

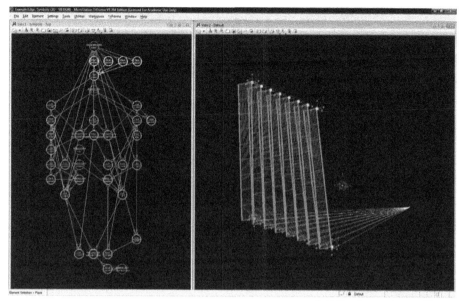

Vertical shades are generated following a similar process that controls variations of vertical overhangs according to their orientation relative to the north direction.

A combination of both the horizontal and vertical features creates the proposed parametric shading system. The design system calculates the states in between the main cordial directions and generates the best fitting lengths of overhangs according to the orientation of the facades.

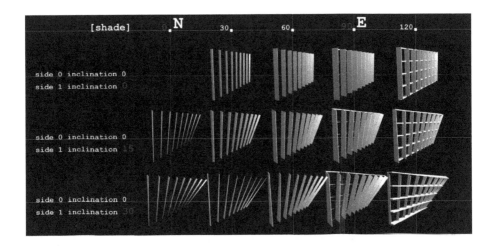

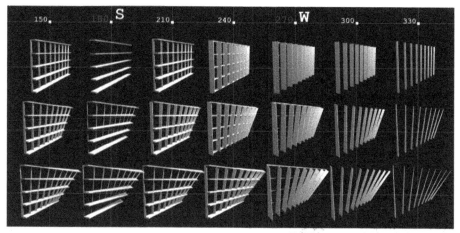

5.15 Instances of the shading system

Figure 5.15 shows the complete shading system applied to a screen wall. Orientation angles are mapped out in 30 degrees increments. In the first row, the diagram shows how the parametric shading system responds to varying orientations by gradually changing from one configuration to the other. Starting with a north-facing wall the shading overhangs transition to a vertical configuration facing east, and then to a horizontal configuration facing south. The second and third rows illustrate the transformation of the shading system on an inclined facade. Because of the geometry of the façade the shading system takes a three dimensional configuration.

Application

Coordinating the shading system with the explorations of the *Tower Formation* exploration illustrates its capacity. Polygons are populated on the exterior of the tower to serve as a base for the application of the sun shading system. The populated shape polygons consist of a series of four sided polygons that vary in dimensions and angles. The polygons form a second skin based on the forming points of the tower. Each of the shape polygons is one floor in height.

According to the varying orientations of each of the tower facades the sunshade design system automatically calculates the relevant combination of horizontal and vertical. The vertical divisions in the system are set to four. As a result, there are three horizontal overhangs on the facade between floors. The vertical spacing of the overhangs is equal to the horizontal spacing.

5.16 *Shade* system application 1

Figure 5.16 shows the system's population over the exterior polygons of the tower. The vertical overhangs (fins) are shown in red, horizontal overhangs are shown in green. The illustration illustrates the response of the shading system. South-facing sides result in additional horizontal shading elements with varying lengths.

144

Rendered images (figures 5.17 to 5.21) illustrate the application of the shading system.

Figure 5.17 shows the northeast elevation of the tower with vertical fins.

On the southeast side of the tower (figures 5.18 and 5.19), horizontal and vertical overhangs respond to the exposure to the sun from the south and east.

5.18 *Shade* system application 3

5.17 *Shade* system application 2

5.19 *Shade* system application 4

145

Figures 5.20 to 5.22 illustrate the varying lengths of overhangs in detail. Figure 5.20 shows a section of a southeast-facing facade with folds. Varying lengths of vertical and horizontal overhangs correspond to shifts in the orientation of the facade. In a similar way, the shading system reacts to both the northwest and southwest orientations in figure 5.21. Figure 5.22 shows the southwest and southeast sides of the tower.

5.21 *Shade* system application 6

5.20 *Shade* system application 5

5.22 *Shade* system application 7

Comments

Shade illustrates the application of a rule-based design approach on the formation of a small-scale sun shading system. The system uses the orientation of screen walls to adjust horizontal and vertical louvers that provide maximum sun protection. The system is modeled using parametric modeling software that allows for a dynamic response of the design system to its changing parameters. Modifications to the orientation of the screen wall results in new calculations of the overhangs of the shading system.

The designer controls the starting form of the components of the design system, the parameters to which the system responds, and the types of responses. However, the parametric system generates the final form of the shading device based of the dynamic relationships between components and influencing parameters.

The rule-based approach to design used for *Shade* can be further developed to accommodate real-time changes in light conditions. Coordinated with light sensors, the design system can be expanded to control adjustable fins and louvers.

6. City Configurations

Introduction

City Configurations explores parametric design as part of a responsive approach to city planning. The study illustrates how parametric modeling can contribute to dynamic relationships between individual building designs and the development of a city. The exploration makes use of parametric modeling to establish relationships between existing and future buildings. It illustrates a parametric system as a planning tool that incorporates planning regulations.

Design Approach

Zoning bylaws are rules set by city planning departments in order to regulate current and future developments in the city. Rule-based design approaches hold great potential in urban planning because direct and explicit laws generally guide the planning process. Parametric modeling is useful for executing multiple procedures and controlling multiple processes. This capacity allows developing complex bylaw and regulations based on dynamic relationships between buildings.

The main focus of this exploration is the integration of planning regulations into a parametric system to create a responsive model that incorporates current and future buildings in the city. This system follows planning rules to regulate and calculate forms, heights, and locations of future buildings. As a result, the use of parametric modeling at the planning scale promotes urban planning that takes into account dynamic relationship between buildings at their surroundings.

Similar to previous explorations, a rule-based design approach is used to produce programmatic and formal complexity out of simple initial states. First, neighbourhood buildings are abstracted into simple geometries and modeled as part of a parametric system. Second, planning regulations and laws are translated into a set of parameters that guide geometries within the parametric system. These rules dictate how the system responds to its changing influences. Third, changes of parameters result in dynamic responses of the design system

Design Background

The City of Vancouver regulates the design of new developments through Land Use and Development Policies and Guidelines. The city divides the downtown area of Vancouver into several districts or neighbourhoods. Each district has its own regulations. The site for this exploration is Vancouver's Downtown South that is also used for as the context for *Tower Formation* in chapter four.

The emphasis of city regulations and design guidelines[vi] in Vancouver is generally on:

1. Public and private views: Views to the mountains in the north are greatly appreciated and protected. Guidelines regarding the preservation of these views from public spaces defined as view corridors regulate the maximum heights of buildings in the downtown area. The maximum allowable height in Downtown South is 300 ft.

The views from private residential units to the mountains are preserved by the general orientation of the downtown street grid. The grid is rotated 45 degrees to the north to allow maximum views of the mountains and to promote general lighting condition in the downtown area. Towers sizes are limited by a maximum floor plate area of 6500 sq.ft. in order to minimise obstruction of views. These regulations generally result in slim tower configurations. (Figure 6.1)

6.1 City of Vancouver guidelines and regulations

2. Light and ventilation: The Downtown South design guidelines focus on minimising shadow impact on parks, public open spaces, and major streets. This results in a required separation of 80 ft. between towers in the city core. Larger separations are encouraged to let natural lighting penetrate between the buildings to open gathering spaces.

3. Continuous street edge definition: The City of Vancouver calls for a consistent pattern in the street edge definition to produce a lively residential community inside the city core. The objective to create active streets with visual interest for pedestrians is promoted by assigning townhouses podiums to each residential tower.

4. Privacy: Privacy in buildings is promoted by regulating setbacks for each site.

Position

Parameters for the parametric system for *City Configurations* are based on the shortcomings of current city planning regulations in the Vancouver Downtown South. Rather than reviewing individual building developments *City Configurations* established relationships between buildings during the planning process.

By deploying the rule-based design approach the study critiques:

1. Fixed planning regulations: The bylaws, regulations, and design guideline that govern the configuration of existing and new buildings are based on fixed rules that are applied to all buildings in neighbourhood. These rules are not designed to consider the relationship between buildings nor the effect they have on each other.

2. Individual project assessment: Currently, each new building project is assessed individually. However, relationships between buildings should be a central aspect of the permit process to consider the effects of neighbouring buildings and to acknowledge the influences of new buildings on their urban context.

Proposal

The objective of *City Formations* is to integrate a range of planning rules and regulations that define the relationship between towers into a parametric design model. The system adjusts the position and configuration of new towers in response to the dynamic state of their surrounding context.

Design process

An analysis of critical planning regulations in Vancouver forms the basis for the parametric design system for *City Formations*. The parametric design system dynamically relates spatial parameters that result the guidelines and regulations and allows for a design process responsive to an urban context in flux.

Criteria:

The proposed system regulates the configuration of buildings based on the following criteria:

1. Fixed density:

Vancouver city planning guidelines set maximum allowable heights and floor plate areas to control the density in the city core. These regulations are applied to all buildings in the particular district regardless of the current and future situation. Instead of applying fixed heights and areas, *City Configurations* considers the desired density for districts inside the city core. A constant building density can be achieved by setting a maximum collective volume of buildings in that particular area.

$$\sum V = \sum (H \times A) = constant$$

If floor plate areas are fixed, then the collective building volume is directly proportional to heights of buildings. Thus, the parametric design system governs the height relationship between towers to control the maximum allowable volume and density. The maximum allowable height is distributed to all buildings in the district instead of regulating the heights per project. Heights of individual towers are dynamically related. If one tower is more than average height, the system changes the configuration of surrounding towers accordingly.

6.2 Fixed density

Rule:

$$\sum H = H1 + H2 + H3 + H4++ = constant$$

This collective approach to regulating heights applies a non-linear change to the maximum allowable heights. In other words, one particular building's height affects its immediate neighbours by changing their allowable heights. Buildings at a greater distance to a tall building are less affected than immediately adjacent towers. (Figure 6.2)

2. Shading Impact:

The design guidelines for Vancouver Downtown South lack specific rules to consider natural lighting and shading. Currently, the City of Vancouver requires a shading analysis as part of a building application to review the shading impact of a proposed building on its surrounding public spaces. Developments over 35 feet in height require a shadow impact analysis taken at the equinox at 10:00 a.m., noon, 2:00 p.m., and 4:00 p.m.

Instead of reviewing the shading analysis for every building in an independent process, *City Configurations* models the shadow analysis together with rules that regulate the effect

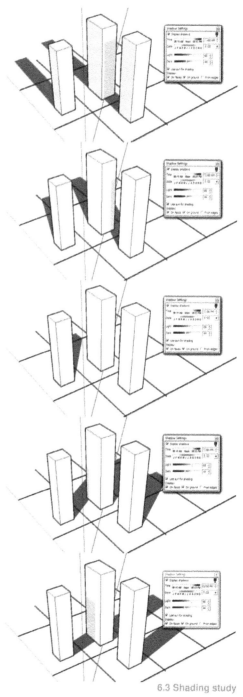

6.3 Shading study

153

of a shadow cast by a building on its context. A shadow cast by an existing building dictates allowable positions of future buildings. The parametric design system establishes a dynamic relationship between building heights, shadows cast, and building locations.

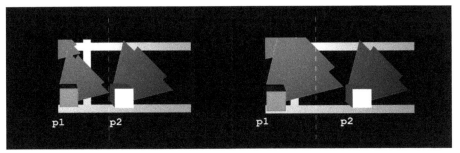

6.4 Shading impact

Rule:

No shadows should be cast on buildings from 10:00 am to 2:00 pm between the March and September equinoxes.

A shadow analysis illustrates the footprint of shadows in relation to heights. (Figure 6.3) Because the shadow of a building is proportional to its height and size, the shadow footprint is used in the parametric system to control the required and allowable heights and separations between buildings. In the dynamic parametric model positions, buildings are set to avoid shadow casting on neighbouring buildings. (Figure 6.4)

p2 = p1 + L shading

This dynamic model responds to changes in heights, position, and footprint of any of the buildings, and accordingly set the allowable configurations of other buildings in the neighbourhood.

3. View Blocking:

As the view to the northern mountains is strongly encouraged in the design guidelines of the City of Vancouver the parametric design system controls the relationship between buildings to avoid blocking views to the north. The system calculates a building's orientation towards north and scans the surrounding buildings. If a proposed positions of a building blocks views from other buildings, the system will automatically reconfigures the allowable floor plate size of the future building to minimise the blocking.

154

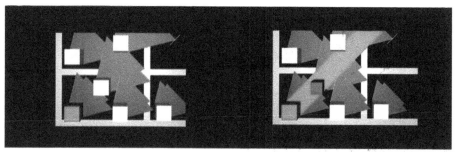

Rules:

1. The blocking of north views should be minimised between towers.
2. Floor plate size of a tower that blocks sight lines should decrease to allow for views.

A view test is based on lines from the face of one building towards the north direction. If the extension lines hit other buildings then the view is obstructed. In this case, the allowable floor plate size for the obstructing buildings is decreased. (Figure 6.5) criteria   start   function

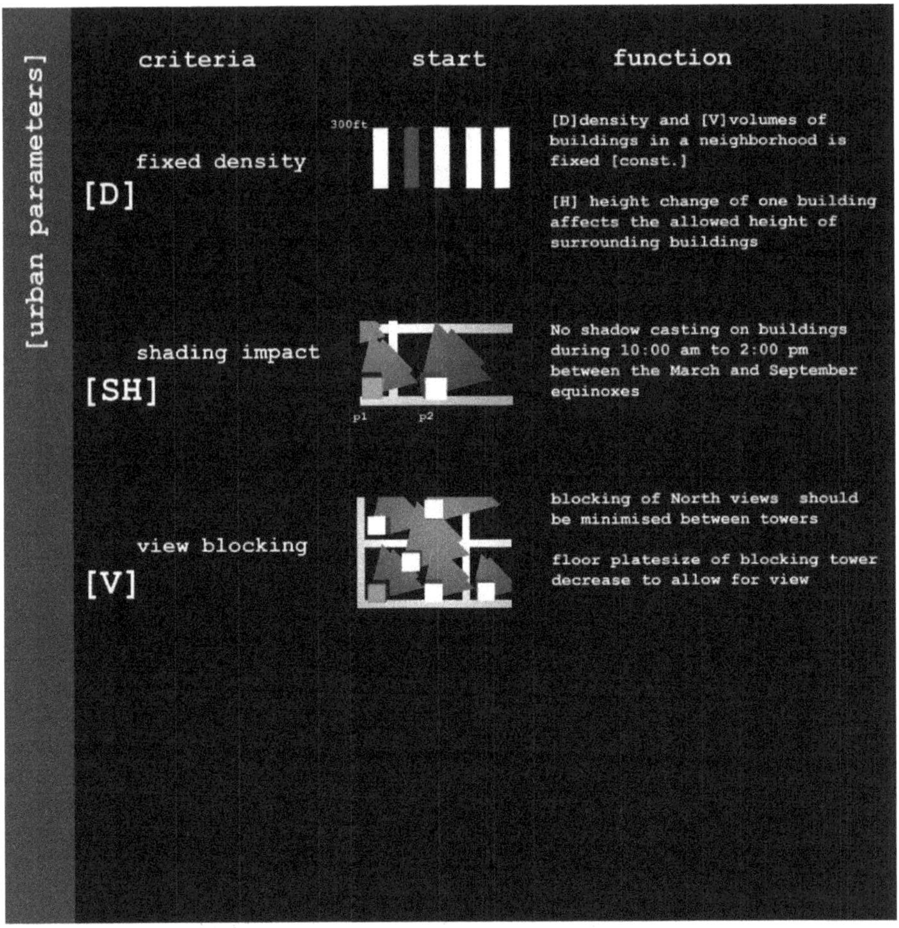

[urban parameters]

fixed density

[D]

300ft

[D]density and [V]volumes of
buildings in a neighborhood is
fixed [const.]

[H] height change of one building
affects the allowed height of
surrounding buildings

shading impact

[SH]

p1   p2

No shadow casting on buildings
during 10:00 am to 2:00 pm
between the March and September
equinoxes

view blocking

[V]

blocking of North views should
be minimised between towers

floor platesize of blocking tower
decrease to allow for view

156

```
control                              input           output

Σ V = Σ (H x  A) = const.       > Variable h_1      300ft
                                > Variable h_2
                                > Variable h_3
                                > Variable h_4
Σ H = H1+ H2+ H3+ H4++  = const.  > Variable h_5
                                > Variable h_6


p2 = p1 + L shading             > Variable h_1
                                > Variable h_2
                                > Variable h_3
                                > Variable h_4
                                > Variable h_5
                                > Variable h_6        p1      p2


transaction script "running VIEW TEST"
{
    if ( projectb5.X < building5_b.XTranslation)
      {
        if ( projectb5.Y > building5_b.YTranslation)
          {
            if ( projectb5.X > building5_a.XTranslation)
              {
                if ( projectb5.Y < building5_a.YTranslation)
                  {
                    b5color = 2;

                    Building5.Xdimension=side-5;
                    Building5.Ydimension=side-5;
                  }
              }
          }
      }
}
```

6.6 *City Configurations* parameters

157

```
transaction modelBased "creating point grid"
{
    feature point07 GC.Point
    {
        CoordinateSystem          = baseCS;
        XTranslation              = Series(0,2*xblock,xblock);
        YTranslation              = Series(0,3*yblock,yblock);
        ZTranslation              = <free> (0.0);
        SymbolXY                  = {104, 100};
        Color                     = 0;
        FillColor                 = -1;
        Free                      = true;
        HandlesVisible            = true;
        IsModifiable              = true;
        Level                     = 0;
        LevelName                 = "Default";
        LineStyle                 = 0;
        LineStyleName             = "0";
        LineWeight                = 0;
        MaximumReplication        = true;
        Replication               = ReplicationOption.AllCombinations;
        SymbologyAndLevelUsage
SymbologyAndLevelUsageOption.AssignToFeature;
        Transparency              = 1;
    }
    feature xblock GC.GraphVariable
    {
        Value                     = 25;
        RangeStepSize             = 0.0;
        SymbolXY                  = {105, 101};
    }
    feature yblock GC.GraphVariable
    {
        Value                     = 10;
        RangeStepSize             = 0.0;
        SymbolXY                  = {105, 100};
    }
}

transaction modelBased "creating 1st solid"
{
    feature csBuilding1 GC.CoordinateSystem
    {
        CoordinateSystem          = baseCS;
        XTranslation              = point07[1][1].XTranslation+10;
        YTranslation              = point07[1][1].YTranslation+2.5;
        ZTranslation              = 0;
        SymbolXY                  = {100, 110};
        Color                     = 0;
        FillColor                 = -1;
        Free                      = true;
        Level                     = 0;
        LevelName                 = "Default";
        LineStyle                 = 0;
        LineStyleName             = "0";
        LineWeight                = 0;
        MaximumReplication        = true;
        SymbologyAndLevelUsage
SymbologyAndLevelUsageOption.AssignToFeature;
        Transparency              = 1;
    }
    feature h1 GC.GraphVariable
    {
        Value                     = 20;
    }
    feature solid01 GC.Solid
    {
        CoordinateSystemAtOrigin  = csBuilding1;
        Xdimension                = 3;
        Ydimension                = 3;
        Zdimension                = h1;
    }
    feature xgrid GC.GraphVariable
    {
        Value                     = 75;
    }
}
```

158

System making

The following illustrates the process of creating the parametric system for *City Configurations* in *GenerativeComponents*. First, context features such as streets and buildings in the downtown district are modeled as basic geometries. The three design criteria regarding density, shading, and view are then integrated as rules and relationships to guide the interaction between the parameters of the parametric design system. Consistent with the rules and constraints that were previously established modifications of any of the parameters results in adjustments of the configuration of the entire model.

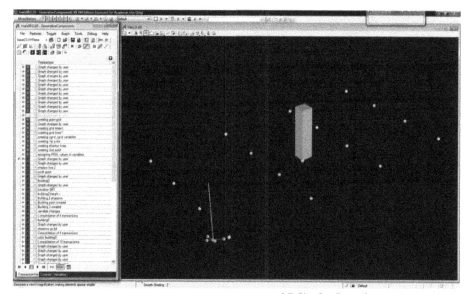

6.7 *City Configurations* system making 1

A point grid *point07* represents street intersections. A solid volume is created to represent the first building in the neighbourhood.

```
transaction modelBased "creating xgrid ygrid variables"
{
    feature point07 GC.Point
    {
        XTranslation            = Series(0,xgrid,xblock);
        YTranslation            = Series(0,ygrid,yblock);
    }
    feature xblock GC.GraphVariable
    {
        LimitValueToRange       = true;
        RangeMaximum            = 50.0;
    }
    feature xgrid GC.GraphVariable
    {
        Value                   = 50;
        RangeStepSize           = 0.0;
    }
    feature yblock GC.GraphVariable
    {
        Value                   = 10.0;
        LimitValueToRange       = true;
        RangeMaximum            = 50.0;
    }
    feature ygrid GC.GraphVariable
    {
        Value                   = 30;
        RangeStepSize           = 0.0;
    }
}
transaction modelBased "Graph changed by user"
{
    feature road GC.GraphVariable
    {
        Value                   = 3;
        RangeStepSize           = 0.0;
    }
}
transaction script "creating grid linesX"
{
    {for (int i=0;i<5;i++)

        {
            Line    line=new   Line("line_"+i).ByPoints(point07[0][i],   point07
[5][i]);
        }
        {
            Line line=new Line("line__0").Offset(line_0, road, baseCS.XZPlane );
        }

        {
            BSplineSurface                              BSplineSurface=new
BSplineSurface("surface__0").Ruled(line__0, line_0 );
        }
        {
            Line line=new Line("line__1").Offset(line_1, road, baseCS.XZPlane );
        }
        {
            BSplineSurface                              BSplineSurface=new
BSplineSurface("surface__1").Ruled(line__1, line_1 );
        }
        {
            Line line=new Line("line__2").Offset(line_2, road, baseCS.XZPlane );
        }
        {
            BSplineSurface                              BSplineSurface=new
BSplineSurface("surface__2").Ruled(line__2, line_2 );
        }
        {
            Line line=new Line("line__3").Offset(line_3, road, baseCS.XZPlane );
        }
        {
            BSplineSurface                              BSplineSurface=new
BSplineSurface("surface__3").Ruled(line__3, line_3 );
        }
        {
            Line line=new Line("line__4").Offset(line_4, road, baseCS.XZPlane );
        }
        {
            BSplineSurface                              BSplineSurface=new
BSplineSurface("surface__4").Ruled(line__4, line_4 );
        }

    }
}
```

160

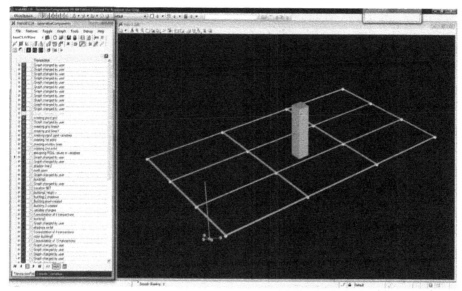

6.8 *City Configurations* system making 2

A series of lines pass through the initial grid point *point07* to define street edges. The variable *Road* controls the width of the streets.

The point grid spacing is adjusted to reflect the actual city block dimensions.

The variables *Xblock*, *Xgrid*, *Yblock*, and *Ygrid* control the block size (increment) and the size of the area in study (end) both in X and Y directions.

161

```
transaction modelBased "creating shadow lines"
{
    feature building1shadowa GC.Line
    {
        StartPoint                    = csBuilding1;
        Direction                     = direction01;
        Length                        = h1;
        SymbolXY                      = {101, 113};
        Color                         = 8;
        FillColor                     = -1;
        Free                          = true;
        Level                         = 0;
        LevelName                     = "Default";
        LineStyle                     = 0;
        LineStyleName                 = "0";
        LineWeight                    = 0;
        MaximumReplication            = true;
        Transparency                  = 1;
    }
feature shadowdirection GC.Direction
    {
        Origin                        = point07[0][0];
        DirectionPoint                = point06;
        SymbolXY                      = {101, 112};
        SymbolicModelDisplay          = null;
        Color                         = 0;
        FillColor                     = -1;
        Free                          = true;
        Level                         = 0;
        LevelName                     = "Default";
        LineStyle                     = 0;
        LineStyleName                 = "0";
        LineWeight                    = 0;
        MaximumReplication            = true;
        PartFamilyName                = null;
        PartName                      = null;
        RoleInExampleGraph            = null;
        Transparency                  = 1;
    }
transaction modelBased "creating 2nd solid"
{
    feature Building GC.Solid
    {
        CoordinateSystemAtOrigin      = csBuilding;
        Xdimension                    = side;
        Ydimension                    = side;
        Zdimension                    = h1;
        SymbolXY                      = {100, 125};
        SymbolicModelDisplay          = null;
        Color                         = color;
        FillColor                     = -1;
        Free                          = true;
        Level                         = 0;
        LevelName                     = "Default";
        LineStyle                     = 0;
        LineStyleName                 = "0";
        LineWeight                    = 0;
        MaximumReplication            = true;
        PartFamilyName                = null;
        PartName                      = null;
        RoleInExampleGraph            = null;
        Transparency                  = 1;
    }
}
```

162

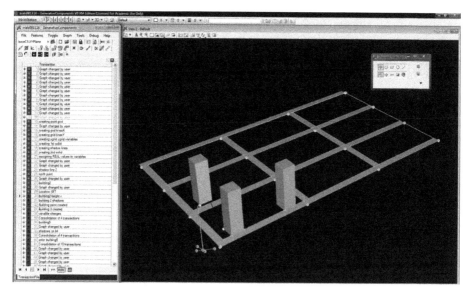

6.9 *City Configurations* system making 3

Buildings 2 and 3 are created and placed in the street grid according to their shading impact. Footprints of shadow casting are modeled for each building.

Building2.Ytranslation = Building1.Ytranslation

Building2.Xtranslation = Building1.Xtranslation + Building1. Side + SL1

Building3.Xtranslation = Building1.Xtranslation

Building3.Ytranslation = Building1.Ytranslation + Building1. Side + SL1

Building1.Side is the side dimension of the 1st building and *SL1* is the length of the 1st building's shadow.

163

```
    transaction modelBased "variable changes"
{
    feature h1 GC.GraphVariable
    {
        Value                   = 70.0;
    }
    feature h2 GC.GraphVariable
    {
        Value                   = 60.0;
    }
    feature h3 GC.GraphVariable
    {
        Value                   = 55.75;
    }
feature building3_X GC.GraphVariable
    {
        LimitValueToRange       = true;
        RangeMinimum            = 20.0;
        SymbolXY                = {86, 105};
    }
    feature building4_X GC.GraphVariable
    {
        Value                   = Abs(h1-(point07[0][1].Y-point10.Y));
        LimitValueToRange       = true;
        RangeMinimum            = 20.0;
        RangeStepSize           = 0.0;
    }
    feature csBuilding GC.CoordinateSystem
    {
        YTranslation            = point07[0][1].Y+building4_X;
        SymbolXY                = {91, 103};
    }
    feature csBuilding1 GC.CoordinateSystem
    {
        YTranslation            = point07[0][0].YTranslation+road+ 10;
        SymbolXY                = {93, 103};
    }
    feature csBuilding2 GC.CoordinateSystem
    {
        XTranslation            = point07[1][0].X-building2_X;
        SymbolXY                = {89, 103};
    }
    feature csBuilding3 GC.CoordinateSystem
    {
        SymbolXY                = {85, 103};
    }
}
```

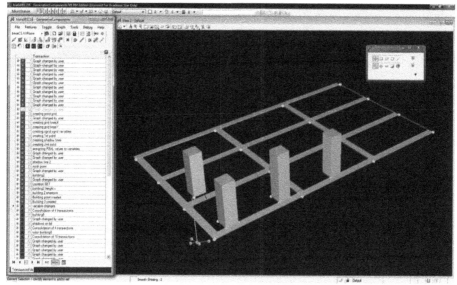

6.10 *City Configurations* system making 4

Building4 is created and positioned relative to *Building2* according to the shading impact rule.

Building4.Ytranslation = Building2.Ytranslation

Building4.Xtranslation = Building2.Xtranslation + Building2. Side + SL2

165

```
transaction modelBased "dynamic height relations sum to const"
{
    feature Building GC.Solid
    {
        Zdimension              = h4+h_var;
    }
    feature Building2 GC.Solid
    {
        Zdimension              = h2+h_var;
    }
    feature Building3 GC.Solid
    {
        Zdimension              = h3+h_var;
    }
    feature Building5 GC.Solid
    {
        Zdimension              = h5+h_var;
    }
    feature Building6 GC.Solid
    {
        Zdimension              = h6+h_var;
    }
    feature h1 GC.GraphVariable
    {
        Value                   = 55.0;
    }
    feature h2 GC.GraphVariable
    {
        Value                   = 80.0;
    }
    feature h3 GC.GraphVariable
    {
        Value                   = 75.0;
    }
    feature h4 GC.GraphVariable
    {
        Value                   = 28.05;
    }
    feature h5 GC.GraphVariable
    {
        Value                   = 60.0;
    }
    feature h6 GC.GraphVariable
    {
        Value                   = 77.0;
    }
    feature h_sum GC.GraphVariable
    {
        Value                   = 460;
        RangeStepSize           = 0.0;
    }
    feature h_var GC.GraphVariable
    {
        Value                   = (h_sum-(h1+h2+h3+h4+h5+h6))/6;
        RangeStepSize           = 0.0;
    }
    feature solid01 GC.Solid
    {
        Zdimension              = h1+h_var;
    }
}
```

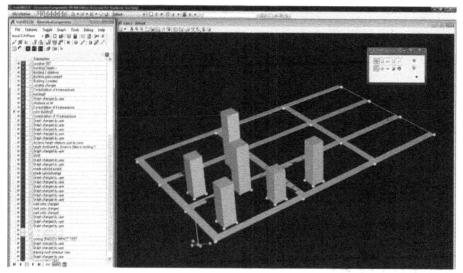

The next buildings *building 5* and *building 6* are created according to the same positioning rules. Height relationships are established between the buildings. Variables *h1*, *h2*, *h3*, *h4*, *h5* and *h6* are the individual heights of each of the buildings. In order to follow the pre-established fixed density rule, their combined heights equals a fixed value (constant). Changes in any of the height of the building result in a non-linear configuration change of other building heights and positions consistent with the height and shadow rules.

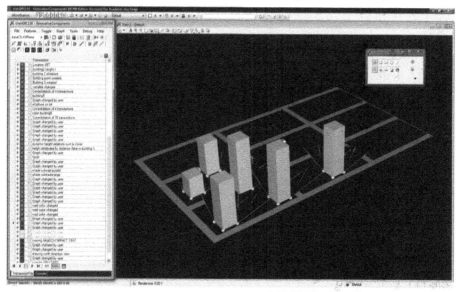

```
transaction modelBased "height distibuted by distance"
{
    feature Building GC.Solid
    {
        Zdimension              = h4+h_var*1.5;
    }
    feature Building3 GC.Solid
    {
        Zdimension              = h3+h_var*.5;
    }
    feature Building6 GC.Solid
    {
        Zdimension              = h6+h_var*.5;
    }
    feature h1 GC.GraphVariable
    {
        Value                   = 75.0;
    }
    feature h3 GC.GraphVariable
    {
        Value                   = 85.0;
    }
    feature h5 GC.GraphVariable
    {
        Value                   = 30.0;
    }
}


transaction script "POSITION TEST"
{
    if ( building5_a.YTranslation > point07[0][1].Y)


                                    {
                                        b5color=1;
                                        color=1;
                                        h1=h1max;

                                    }

                    if ( building6_a.YTranslation > point07[0][2].Y)


                                    {
                                        color=3;


                                    }
}
```

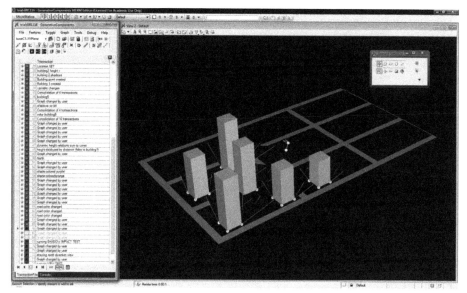

6.13 *City Configurations* system making 7

A positioning test reconfigures a building and the maximum allowable heights of other buildings if an initial model configuration violates positioning rules. In this illustration, the initial position of *building 5* violates street clearance. As a result, the system adjusts its position and therefore, the maximum allowable height for *building1*.

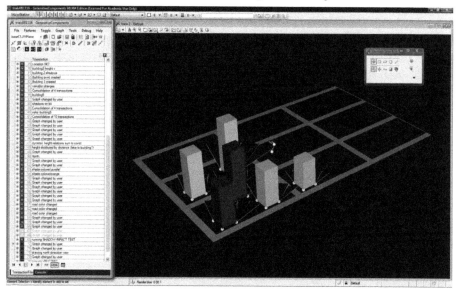

6.14 *City Configurations* system making 8

```
transaction script "running VIEW TEST"
{
    if ( projectb5.X < building5_b.XTranslation)
        {
            if ( projectb5.Y > building5_b.YTranslation)
                {
                    if ( projectb5.X > building5_a.XTranslation)
                        {
                            if          (          projectb5.Y          <
building5_a.YTranslation)
                                {
                                    b5color = 1;

                                    Building5.Xdimension=side-6;
                                    Building5.Ydimension=side-6;
                                }
                        }
                }
        }
}
```

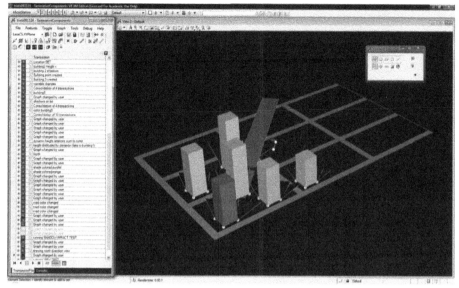

6.15 *City Configurations* system making 9

A view-blocking test projects lines towards the north direction. If a building blocks views towards north from the floor plate area is decreased to minimise the impact of the building.

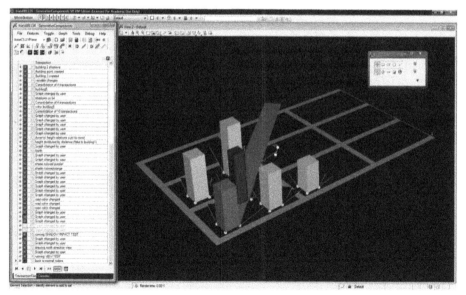

6.16 *City Configurations* system making 10

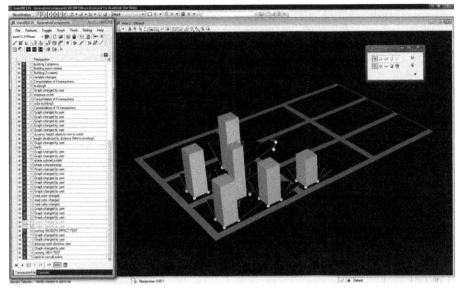

6.17 *City Configurations* system making 11

Figure 6.17 illustrates an instance of a height and positioning configuration consistent with the rules of height, shading impact, and view blocking.

Comments

City Configurations illustrates a method to establish a dynamic approach to city planning. A parametric design system regulates the existing and future relationships of buildings by controlling allowable locations and configurations. A parametric model similar to *City Configurations* can be used in city planning to map existing buildings in a particular neighbourhood and to coordinate and illustrate planning guidelines for existing and new buildings.

In principal, the parametric design system can be expanded to include criteria like costs and property values using a similar logic.

[vi] http://vancouver.ca/commsvcs/guidelines/pol&guide.htm#dd

Screen clipping taken: 28/10/2008, 4:32 PM

7. Conclusion

Emergent Programmatic Form-ation applies parametric design techniques to a series of explorations at scales varying from living spaces to urban scale configurations. The objective of the study is to establish a responsive process of design and regulation; programmatic and context-sensitive parameters set rules and govern the process of design.

This study develops a rule-based approach to design through the understanding of emergence and its potential to solve complex problems, such as programmatic concerns in designs that reflect and respond to larger sets of interrelated social and environmental references. The consecutive design explorations of *Emergent Programmatic Form-ation* engage in increasing complexity. All four explorations of this study can be coordinated: *Space Configurations* and *Shade* can become an integral part of the *Tower Formation*, The exploration of tower configurations can in turn be integrated into *City Configurations* to achieve added complexity and detail.

It is important to note the explorations of this study do not propose finalised design solutions but intend to illustrate a method to approach complex design problems in computational environments. The study compares design methods using parametric tools to conventional approaches to form making. Parametric modeling allows addressing a complex set of social and programmatic criteria in a responsive and interactive design setting at a variety of scales. Previous digital techniques used by architects often focus on the generation of formal complexity without reflecting the outside references essential for creating architecture.

A key aspect of the explorations of *Emergent Programmatic Form-ation* is the selection of rules and the translation of design problems and their references into components, parameters, and dependencies inside a parametric model. In principal, if appropriate parameters are created, parametric design techniques can address and coordinate various design and planning concerns and references. Incorporating an extended set of parameters is consistent with the design approach explored in this study.

Parametric tools are based on a complex method of storing explicit design rules. Unlike with CAD modeling, the investment is not so much in the geometries created, but in the rules, relationships, and dependencies that define how geometries interact and respond to changing references.

While based on rules, parametric design systems rely on the judgment of the designer. Exclusions, limits, and constraints have to be defined in order to set the allowable range of system responses. These rules are explicit and defined prior to starting a design.

174

The selection of references can involve a range of participants, thereby opening up the design process to input of collaborative design teams, clients, and users. The collective input can be considered in the configuration of parametric design systems, which contributes to move the design process beyond the creation of form to incorporate a broader range of factors.

> *"The architectural field's current use of the parametric has been superficial and skin-deep, may be importantly so, lacking of a larger framework of referents, narratives, history, and forces."* (Michael Meredith 6)

By focusing on a variety of scales, diverse user groups, and planning regulations that reflect the cultural context for design in a city, the explorations of *Emergent Programmatic Form-ation* illustrate how parametric design can be used to move beyond the development complex formal solutions. The study illustrates the use of parametric design techniques in designs that are responsive to the particular conditions of context, user functions, and program based on a complex set of interrelated natural and social references specific to urban housing.

Figures credit

All figures are produced by Yehia Madkour, unless indicated below:

Figure 2.1 Son O House
Spuybroek, Lars. NOX. Thames & Hudson, 2004.

Figure 2.2 Son O House Process
Spuybroek, Lars. NOX. Thames & Hudson, 2004.

Figure 2.3 Log Cabin Process 1
Aranda, Benjamin and Chris Lasch. Tooling. New York: Princeton Architectural Press, 2005,23,25.

Figure 2.4 Log Cabin Process 2
Aranda, Benjamin and Chris Lasch. Tooling. New York: Princeton Architectural Press, 2005,29,30.

Figure 2.5 Pneumatic Strawberry Bar Process
http://www.achimmenges.net/achimmenges_pl04_DesignExper.swf
(Accessed 13 4 09)

Bibliography

aadrl.net. 2 4 2009 <http://www.aadrl.net/>.

Aish, Robert. "Positioning of GenerativeComponents." GenerativeComponents Version 08.09.04.76 Help . 2007.

Benjamin Aranda, chris Lasch. Tooling. New York: Princton Architectural Press, 2005.

Bernard Tschumi, Irene Cheng. The State of Architecture at the beginning of the 21st century. New York: The Monacelli Press, 2003.

Castle, Helen. "Emergence in Architecture." AA Files No 50 Spring 2004: 50-61.

Chu, Karl. "Metaphysics of Genetic Architecture and Computation." AD Wiley Academy Vol.76 No.4 2006: 38-45.

City of Vancouver, Planning Department, Statistics and Information. 28 10 2008. <http://vancouver.ca/commsvcs/planning/stats.htm>.

DeLanda, Manuel. "Deleuze and the Use of Genetic Algorithms in Architecture." AD Wiley Academy January 2002: 9-12.

Gilles Deleuze, Felix Guattari. A Thousand Plateaus. University of Minnesota Press, 1987.

Hensel, Michael. Emergence: Morphogenetic Design Strategies. Academy Press, 2004.

Holland, John. Emergence: From Chaos to Order. Basic Books, 1998.

Landa, Manuel De. A Thousand Years of Nonlinear History. New York: Zone Books, 2000.

Leach, Neil. "Digital Morphogenesis: A New Paradigmatic Shift in Architecture." Archithese 2006: 44-49.

Menges, Achim. http://www.achimmenges.net/. 13 4 2009.

—. "Morpho-Ecologies: Approaching Complex Environments." AD Wiley Academy Vol.74 No.3 2004: 80-89.

Michael Hensel, Achim Menges, Michael Weinstock. "Emergence in Architecture." AD Wiley Academy Vol.74 No.3 2004.

—. Emergence: Morphogenetic Design Strategies. AD Wiley Academy, 2004.

—. "Frei Otto in Conversation with the Emergence and Design Group." AD Wiley Academy Vol.74 No.3 2004: 18-25.

Michael Meredith, Mutsuro Sasaki, Aranda-Lasch. From Control to design: Parametric/Algorithmic Architecture. Actar, 2008.

Michael Weinstock, Achim Menges, Michael Hensel. "Fit Fabric: Versatility through Redundancy and Differentiation." AD Wiley Academy Vol.74 No.3 2004: 40-47.

Michel Hensel, Achim menges, Michael Weinstock. "Techniques and Technologies in Morphogenetic Design." AD Wiley Academy Vol.74 No.3 2004.

Picon, Antoine. "Architecture and the Virtual towards New Materiality." Praxis 6: New Technologies/ New Architecture 2004: 114-121.

—. "Toward a Well-Tempered Digital Design." Harvard Design Magazine No.25 Fall 2006/ Winter 2007: 77-83.

Rahim, Ali. Catalytic Formations: Architecture and Digital Design. Taylor & Francis, 2006.

—. Contemporay Techniques. AD wiley Academy, 2002.

Rocker, Ingeborg M. "When Code matters." AD Wiley Academy Vol.76 No.4 2006: 16-25.

SmartGeometry. 15 2 2007 <http://www.smartgeometry.org/>.

Spuybroek, Lars. NOX. Thames & Hudson, 2004.

Terzidis, Kostas. Algorithmic Architecture. Architectural Press, 2006.

Thompson, D'Arcy Wentworth. On Growth and Form. Cambridge University Press, 1942.

Una May O'Reilly, Martin Hemberg, Achim Menges. "Evolutionary Computation and Artificial Life in Architecture: Exploring the potential of Generative and Genetic Algorithms as Operative Design Tools." AD Wiley Academy Vol.74 No.5 2004: 48-53.

Vancouver Views: Council Approved View Cones. 18 10 2008 <http://vancouver.ca/commsvcs/views/viewcones/91.htm>.

Weinstock, Michael. "Morphogenesis and the Mathematics of Emergence." AD Wiley Academy Vol.74 No.3 2004: 10-17.

www.ingramcontent.com/pod-product-compliance
Lightning Source LLC
LaVergne TN
LVHW042335060326
832902LV00006B/175